Contents

Introduction

The aim of this Guide is to help people find their way around the night sky, by showing how the stars that are visible change from month to month and by including details of various events that occur throughout the year. The objects and events described may be observed with the naked eye, or nothing more complicated than a pair of binoculars.

The conditions for observing naturally vary over the course of the year. During the summer, twilight may persist throughout the night and make it difficult to see the faintest stars. There are three recognized stages of twilight: civil twilight, when the Sun is less than 6° below the horizon; nautical twilight, when the Sun is between 6° and 12° below the horizon; and astronomical twilight, when the Sun is between 12° and 18° below the horizon. Full darkness occurs only when the Sun is more than 18° below the horizon. During nautical twilight, only the very brightest stars are visible. During astronomical twilight, the faintest stars visible to the naked eye may be seen directly overhead, but are lost at lower altitudes. As the diagram shows, during most of June full darkness never occurs at the latitude of Vancouver, BC. Slightly farther south (at Seattle, for example) it is truly dark for about two hours, and for somewhere like Houston, TX, there are at least six hours of darkness.

Another factor that affects the visibility of objects is the amount of moonlight in the sky. At Full Moon, it may be very difficult to see some of the fainter stars and objects, and even when the Moon is at a smaller phase it may seriously interfere with visibility if it is near the stars or planets in which you are interested. A full lunar calendar is given for each month and may be used to see when nights are likely to be darkest and best for observation.

The celestial sphere

All the objects in the sky (including the Sun, Moon and stars) appear to lie at some indeterminate distance on a large sphere, centered on the Earth. This *celestial sphere* has various reference points and features that are related to those of the Earth. If the Earth's rotational axis is extended, for example, it points to the North and South Celestial Poles, which are thus in line with the North and South Poles on Earth. Similarly, the *celestial*

The duration of twilight throughout the year at Vancouver and Houston.

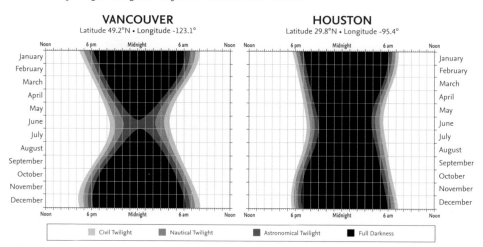

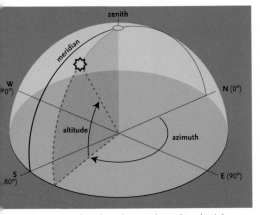

Measuring altitude and azimuth on the celestial sphere.

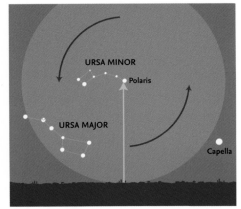

The altitude of the North Celestial Pole equals the observer's latitude.

equator lies in the same plane as the Earth's equator, and divides the sky into northern and southern hemispheres. Because this Guide is written for use in North America, the area of the sky that it describes includes the whole of the northern celestial hemisphere and those portions of the southern that become visible at different times of the year. Stars in the far south, however, remain invisible throughout the year, and are not included.

It is useful to know some of the special terms for various parts of the sky. As seen by an observer, half of the celestial sphere is invisible, below the horizon. The point directly overhead is known as the **zenith**, and the (invisible) one below one's feet as the **nadir**. The line running from the north point on the horizon, up through the zenith and then down to the south point is the **meridian**. This is an important invisible line in the sky because objects are highest in the sky, and thus easiest to see, when they cross the meridian in the south. Objects are said to **transit** when they cross this line in the sky.

In this book, reference is frequently made in the text and in the diagrams to the standard compass points around the horizon. The position of any object in the sky may be described by its **altitude** (measured in degrees above the horizon), and its **azimuth** (measured in degrees from north 0°, through east 90°, south 180° and west 270°). Experienced amateurs and professional astronomers also use another system of specifying locations on the celestial sphere, but that need not concern us here, where the simpler method will suffice.

The celestial sphere appears to rotate about an invisible axis, running between the North and South Celestial Poles. The location (i.e., the altitude) of the Celestial Poles depends entirely on the observer's position on Earth or, more specifically, their latitude. The charts in this book are produced for the latitude of 40°N, so the North Celestial Pole (NCP) is 40° above the northern horizon. The fact that the NCP is fixed relative to the horizon means that all the stars within 40° of the pole are always above the horizon and may, therefore, always be seen at night, regardless of the time of year. This northern circumpolar region is an ideal place to begin learning the sky, and ways to identify the circumpolar stars and constellations will be described shortly.

The ecliptic and the zodiac

Another important line on the celestial sphere is the Sun's apparent path against the background stars – in reality the result of the Earth's orbit around the Sun. This is known as the **ecliptic**. The point where the Sun,

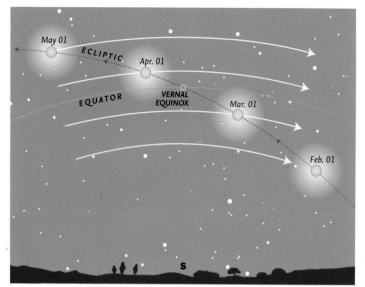

The Sun crossing the celestial equator in spring.

apparently moving along the ecliptic, crosses the celestial equator from south to north is known as the vernal (or spring) equinox, which occurs on March 20. At this time (and at the autumnal equinox, on September 22 or 23, when the Sun crosses the celestial equator from north to south) day and night are almost exactly equal in length. (There is a slight difference, but that need not concern us here.) The vernal equinox is currently located in the constellation of Pisces, and is important in astronomy because it defines the zero point for a system of celestial coordinates, which is, however, not used in this Guide.

The Moon and planets are to be found in a band of sky that extends 8° on either side of the ecliptic. This is because the orbits of the Moon and planets are inclined at various angles to the ecliptic (i.e., to the plane of the Earth's orbit). This band of sky is known as the zodiac and, when originally devised, consisted of 12 **constellations**, all of which were considered to be exactly 30° wide. When the constellation boundaries were formally established by the International Astronomical Union in 1930, the exact extent of most constellations was altered and, nowadays,

the ecliptic passes through 13 constellations. Because of the boundary changes, the Moon and planets may actually pass through several other constellations that are adjacent to the original 12.

The constellations

Since ancient times, the celestial sphere has been divided into various constellations, most dating back to antiquity and usually associated with certain myths or legendary people and animals. Nowadays, the boundaries of the constellations have been fixed by international agreement and their names (in Latin) are largely derived from Greek or Roman originals. Some of the names of the most prominent stars are of Greek or Roman origin, but many are derived from Arabic names. Many bright stars have no individual names and, for many years, stars were identified by terms such as "the star in Hercules' right foot." A more sensible scheme was introduced by the German astronomer Johannes Bayer in the early 17th century. Following his scheme – which is still used today – most of the brightest stars are identified by a Greek letter followed by the genitive form of the constellation's Latin name. An example is the Pole Star, also known as Polaris and α Ursae Minoris (abbreviated α UMi). The Greek alphabet is shown on page 111 with a list of all the constellations that may be seen from latitude 40°N, together with abbreviations, their genitive forms and English names on page 110. Other naming schemes exist for fainter stars, but are not used in this book.

Asterisms

Apart from the constellations (88 of which cover the whole sky), certain groups of stars, which may form a part of a larger constellation or cross several constellations, are readily recognizable and have been given individual names. These groups are known as *asterisms*, and the most famous (and well-known) is the "Big Dipper," the common name for the seven brightest stars in the constellation of Ursa Major, the Great Bear. The names and details of some asterisms mentioned in this book are given in the list on page 111.

Magnitudes

The brightness of a star, planet or other body is frequently given in magnitudes (mag.). This is a mathematically defined scale where larger numbers indicate a fainter object. The scale extends beyond the zero point to negative numbers for very bright objects. (Sirius, the brightest star in the sky is mag. -1.4.) Most observers are able to see stars to about mag. 6, under very clear skies.

The Moon

Although the daily rotation of the Earth carries the sky from east to west, the Moon gradually moves eastwards by approximately its diameter (about half a degree) in an hour. Normally, in its orbit around the Earth, the Moon passes above or below the direct line between Earth and Sun (at New Moon) or outside the area obscured by the Earth's shadow (at Full Moon). Occasionally, however, the three bodies are more or less perfectly aligned to give an *eclipse*: a solar eclipse at New Moon or a lunar eclipse at Full Moon. Depending on the exact circumstances, a solar eclipse may be merely partial (when the Moon does not cover the whole of the Sun's disk), annular (when the Moon is too far from Earth in its orbit to appear large enough to hide the whole of the Sun), or total. Total and annular eclipses are visible from very restricted areas of the Earth, but partial eclipses are normally visible over a wider area.

Somewhat similarly, at a lunar eclipse, the Moon may pass through the outer zone of the Earth's shadow, the **penumbra** (in a penumbral eclipse, which is not generally perceptible to the naked eye), so that just part of the Moon is within the darkest part of the Earth's shadow, the **umbra** (in a partial eclipse); or completely within the umbra (in a total eclipse). Unlike solar eclipses, lunar eclipses are visible from large areas of the Earth.

Occasionally, as it moves across the sky, the Moon passes between the Earth and individual planets or distant stars, giving rise to an **occultation**. As with solar eclipses, such occultations are visible from restricted areas of the world.

The planets

Because the planets are always moving against the background stars, they are treated in some detail in the monthly pages and information is given when they are close to other planets, the Moon or any of five bright stars that lie near the ecliptic. Such events are known as **appulses** or, more frequently, as **conjunctions**. (There are technical differences in the way these terms are defined and should be used in astronomy, but these need not concern us here.) The positions of the planets are shown for every month on a special chart of the ecliptic.

The term conjunction is also used when a planet is either directly behind or in front of the Sun, as seen from Earth. (Under normal circumstances it will then be invisible.) The conditions of most favorable visibility depend on whether the planet is one of the two known as **inferior planets** (Mercury and Venus) or one of the three **superior planets** (Mars, Jupiter and

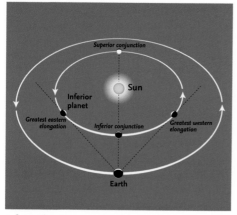

Inferior planet.

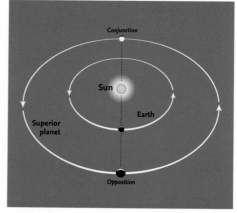

Superior planet.

Saturn) that are covered in detail. (Some details of the fainter superior planets, Uranus and Neptune, are included in this Guide, and special charts for both are given on pages 25.)

The inferior planets are most readily seen at eastern or western **elongation**, when their angular distance from the Sun is greatest. For superior planets, they are best seen at **opposition**, when they are directly opposite the Sun in the sky, and cross the meridian at local midnight.

It is often useful to be able to estimate angles on the sky, and approximate values may be obtained by holding one hand at arm's length. The various angles are shown in the diagram, together with the separations of the various stars in the Big Dipper.

Meteors

At some time or other, nearly everyone has seen a **meteor** – a "shooting star" – as it flashed across the sky. The particles that cause meteors – known technically as "meteoroids" – range in size from that of a grain of sand (or even smaller) to the size of a pea. On any night of the year there are occasional meteors, known as **sporadics**, that may travel in any direction. These occur at a rate that is normally between three and eight in an hour. Far more important, however, are **meteor showers**, which occur at fixed periods of the year, when the

Earth encounters a trail of particles left behind by a comet or, very occasionally, by a minor planet (asteroid). Meteors always appear to diverge from a single point on the sky, known as the **radiant**, and the radiants of major showers are shown on the charts. Meteors that come from a circular area 8° in diameter around the radiant are classed as belonging to the particular shower. All others that do not come from that area are sporadics (or, occasionally from another shower that is active at the same time). A list of the major meteor showers is given on page 31.

Although the positions of the various shower radiants are shown on the charts, looking directly at the radiant is not the most effective way of seeing meteors. They are most likely to be noticed if one is looking about 40°–45° away from the radiant position. (This is approximately two hand-spans as shown in the diagram for measuring angles.)

Other objects

Certain other objects may be seen with the naked eye under good conditions. Some were given names in antiquity – Praesepe is one example – but many are known by what are called "Messier numbers," the numbers in a catalogue of nebulous objects compiled by Charles Messier in the late 18th century. Some, such as the Andromeda Galaxy, M31, and the Orion Nebula, M42, may be seen

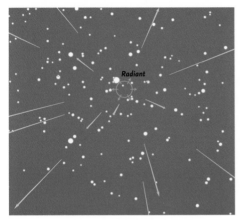

Meteor shower (showing the April Lyrid radiant).

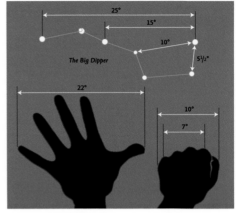

Measuring angles in the sky.

by the naked eye, but all those given in the list will benefit from the use of binoculars. Apart from galaxies, such as M31, which contain thousands of millions of stars, there are also two types of cluster: open clusters, such as M45, the Pleiades, which may consist of a few dozen to some hundreds of stars; and globular clusters, such as M13 in Hercules, which are spherical concentrations of many thousands of stars. One or two gaseous nebulae, consisting of gas illuminated by stars within them, are also visible. The Orion Nebula, M42, is one, and is illuminated by the group of four stars, known as the Trapezium, which may be seen within it by using a good pair of binoculars.

Some interesting objects.

Messier / NGC	Name	Type	Constellation	Maps (months)
—	Hyades	open cluster	Taurus	Sep. – Mar.
—	Double Cluster	open cluster	Perseus	All year
—	Melotte 111 (Coma Cluster)	open cluster	Coma Berenices	Jan. – Aug.
M3	—	globular cluster	Canes Venatici	Jan. – Sep.
M4	—	globular cluster	Scorpius	May – Aug.
M8	Lagoon Nebula	gaseous nebula	Sagittarius	Jun. – Sep.
M11	Wild Duck Cluster	open cluster	Scutum	May – Oct.
M13	Hercules Cluster	globular cluster	Hercules	Feb. – Nov.
M15	—	globular cluster	Pegasus	Jun. – Dec.
M22	—	globular cluster	Sagittarius	Jun. – Sep.
M27	Dumbbell Nebula	planetary nebula	Vulpecula	May – Dec.
M31	Andromeda Galaxy	galaxy	Andromeda	Jun. – Mar.
M35	—	open cluster	Gemini	Oct. – May
M42	Orion Nebula	gaseous nebula	Orion	Nov. – Mar.
M44	Praesepe	open cluster	Cancer	Nov. – Jun.
M45	Pleiades	open cluster	Taurus	Aug. – Apr.
M57	Ring Nebula	planetary nebula	Lyra	Apr. – Dec.
M67	—	open cluster	Cancer	Dec. – May
NGC 752	—	open cluster	Andromeda	Jul. – Mar.
NGC 3242	Ghost of Jupiter	planetary nebula	Hydra	Feb. – May

The Northern Circumpolar Constellations

The northern circumpolar stars are the key to starting to identify the constellations. For anyone in the northern hemisphere they are visible at any time of the year, and nearly everyone is familiar with the seven stars of the Big Dipper: an asterism that forms part of the large constellation of **Ursa Major** (the Great Bear).

Ursa Major

Because of the movement of the stars caused by the passage of the seasons, Ursa Major lies in different parts of the evening sky at different periods of the year. The diagram below shows its position for the four main seasons. The seven stars of the Big Dipper remain visible throughout the year anywhere north of latitude 40°N. Even at the latitude (40°N) for which the charts in this book are drawn, many of the stars in the southern portion of the constellation of Ursa Major are hidden below the horizon for part of the year or (particularly in late summer) cannot be seen late in the night.

Polaris and Ursa Minor

The two stars **Dubhe** and **Merak** (α and β Ursae Majoris, respectively), farthest from the "tail"

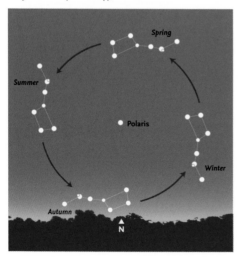

are known as the "Pointers." A line from Merak to Dubhe, extended about five times their separation, leads to the Pole Star, **Polaris**, or α Ursae Minoris. All the stars in the northern sky appear to rotate around it. There are five main stars in the constellation of **Ursa Minor**, and the two farthest from the Pole, **Kochab** and **Pherkad** (β and γ Ursae Minoris, respectively), are known as "The Guards."

Cassiopeia

On the opposite of the North Pole from Ursa Major lies **Cassiopeia**. It is highly distinctive, appearing as five stars forming a letter "W" or "M" depending on its orientation. Provided the sky is reasonably clear of clouds, you will nearly always be able to see either Ursa Major or Cassiopeia, and thus be able to orientate yourself on the sky.

To find Cassiopeia, start with **Alioth** (ε Ursae Majoris), the first star in the tail of the Great Bear. A line from this star extended through Polaris points directly toward γ Cassiopeiae, the central star of the five.

Cepheus

Although the constellation of **Cepheus** is fully circumpolar, it is not nearly as well-known as Ursa Major, Ursa Minor or Cassiopeia, partly because its stars are fainter. Its shape is rather like the end of a house with a pointed roof. The line from the Pointers through Polaris, if extended, leads to **Errai** (γ Cephei) at the "top" of the "roof." The brightest star, **Alderamin** (α Cephei) lies in the Milky Way region, at the "bottom right-hand corner" of the figure.

Draco

The constellation of **Draco** consists of a quadrilateral of stars, known as the "Head of Draco" (and also the "Lozenge"), and a long chain of stars forming the neck and body of the dragon. To find the Head of Draco, locate the two stars **Phecda** and **Megrez** (γ and δ Ursae Majoris) in the Big Dipper,

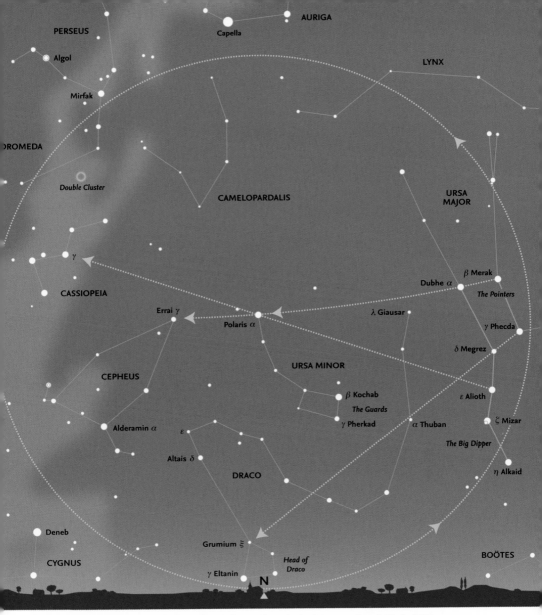

The stars and constellations inside the circle are always above the horizon, seen from latitude 40°N.

opposite the Pointers. Extend a line from Phecda through Megrez by about eight times their separation, right across the sky below the Guards in Ursa Minor, ending at **Grumium** (ξ Draconis) at one corner of the quadrilateral. The brightest star, **Eltanin** (γ Draconis) lies farther to the south. From the head of Draco, the constellation first runs northeast to **Altais** (δ Draconis) and ε Draconis, then doubles back southwards before winding its way through **Thuban** (α Draconis) before ending at **Giausar** (λ Draconis) between the Pointers and Polaris.

The Winter Constellations

The winter sky is dominated by several bright stars and distinctive constellations. The most conspicuous constellation is **Orion**, the main body of which has an hour-glass shape. It straddles the celestial equator and is thus visible from anywhere in the world. The three stars that form the "Belt" of Orion point down towards the southeast and to **Sirius** (α Canis Majoris), the brightest star in the sky. **Mintaka** (δ Orionis), the star at the northeastern end of the Belt, farthest from Sirius, actually lies just slightly south of the celestial equator.

A line from **Bellatrix** (γ Orionis) at the "top right-hand corner" of Orion, through **Aldebaran** (α Tauri), past the "V" of the Hyades cluster, points to the distinctive cluster of bright blue stars known as the **Pleiades**, or the "Seven Sisters". Aldebaran is one of the five bright stars that may sometimes be occulted (hidden) by the Moon. Another line from Bellatrix, through **Betelgeuse** (α Orionis), if carried right across the sky, points to the constellation of **Leo**, a prominent constellation in the spring sky.

Six bright stars in six different constellations: **Capella** (α Aurigae), **Aldebaran** (α Tauri), **Rigel** (β Orionis), **Sirius** (α Canis Majoris), **Procyon** (α Canis Minoris) and **Pollux** (β Gemini) form what is sometimes known as the "Winter Hexagon". Pollux is accompanied to the northwest by the slightly fainter star of **Castor** (α Gemini), the second "Twin".

In a counterpart to the famous "Summer Triangle", an almost perfect equilateral triangle, the "Winter Triangle", is formed by Betelgeuse (α Orionis), Sirius (α Canis Majoris) and Procyon (α Canis Minoris).

Several of the stars in this region of the sky show distinctive tints: Betelgeuse (α Orionis) is reddish, Aldebaran (α Tauri) is orange and Rigel (β Orionis) is blue-white.

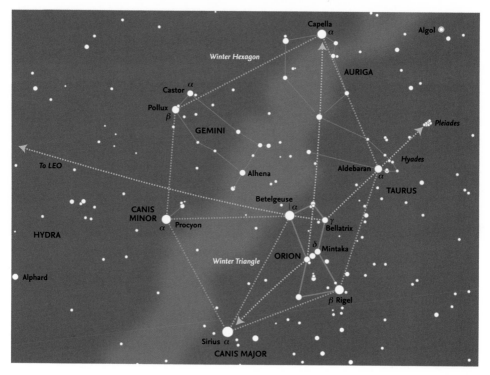

The Spring Constellations

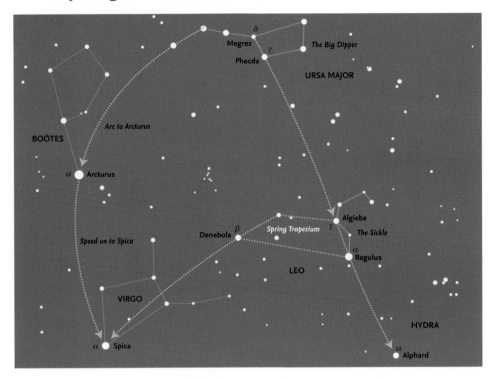

The most prominent constellation in the spring sky is the zodiacal constellation of **Leo**, and its brightest star, **Regulus** (α Leonis), which may be found by extending a line from Megrez and Phecda (δ and γ Ursae Majoris, respectively) – the two stars on the opposite side of the bowl of the Big Dipper from the Pointers – down to the southeast. Regulus forms the "dot" of the "backward question mark" known as "the Sickle". Regulus, like Aldebaran in Taurus is one of the bright stars that lie close to the ecliptic, and that are occasionally occulted by the Moon. The same line from Ursa Major to Regulus, if continued, leads to **Alphard** (α Hydrae), the brightest star in **Hydra**, the largest of the 88 constellations.

The shape formed by the body of Leo is sometimes known as the "Spring Trapezium". At the other end of the constellation from Regulus is **Denebola** (β Leonis), and the line forming the back of the constellation through Denebola points to the bright star **Spica** (α Virginis) in the constellation of **Virgo**. A saying that helps to locate Spica is well-known to astronomers: "Arc to Arcturus and then speed on to Spica." This suggests following the arc of the tail of Ursa Major to Arcturus and then on to Spica. **Arcturus** (α Boötis) is actually the brightest star in the northern hemisphere of the sky. (Although other stars, such as Sirius, are brighter, they are all in the southern hemisphere.) Overall, the constellation of **Boötes** is sometimes described as "kite-shaped" or "shaped like the letter P".

Although Spica is the brightest star in the zodiacal constellation of Virgo, the rest of the constellation is not particularly distinct, consisting of a rough quadrilateral of moderately bright stars and some fainter lines of stars extending outwards.

The Summer Constellations

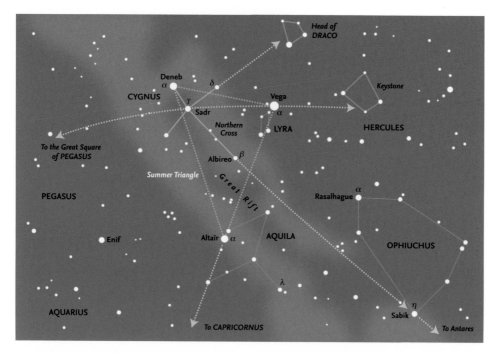

On summer nights, the three bright stars **Deneb** (α Cygni), **Vega** (α Lyrae) and **Altair** (α Aquilae) form the striking "Summer Triangle". The constellations of **Cygnus** (the Swan) and **Aquila** (the Eagle) represent birds "flying" down the length of the Milky Way. This part of the Milky Way contains the **Great Rift**, an elongated dark region, where the light from distant stars is obscured by intervening dust. The dark Rift is clearly visible even to the naked eye.

The most prominent stars of Cygnus are sometimes known as the "Northern Cross" (as a counterpart to the "Southern Cross" – the constellation of Crux – in the southern hemisphere). The central line of Cygnus through **Albireo** (β Cygni), extended well to the southwest, points to **Sabik** (η Ophiuchi) in the large, sprawling constellation of Ophiuchus (the Serpent Bearer) and beyond to **Antares** (α Scorpii) in the constellation of **Scorpius**. Like Cepheus, the shape of Ophiuchus somewhat resembles the gable-end of a house, and the

brightest star **Rasalhague** (α Ophiuchi) is at the "apex" of the "gable".

A line from the central star of Cygnus, **Sadr** (γ Cygni) through δ Cygni, in the northwestern "wing" points towards the Head of Draco, and is another way of locating that part of the constellation. Another line from Sadr to Vega indicates the central portion, "The Keystone", of the constellation of **Hercules**. An arc through the same stars, in the opposite direction, points towards the constellation of **Pegasus**, and more specifically to the "Great Square of Pegasus".

Aquila is less conspicuous than Cygnus and consists of a diamond shape of stars, representing the body and wings of the eagle, together with a rather faint star, λ Aquilae, marking the "head". **Lyra** (the Lyre) mainly consists of Vega (α Lyrae) and a small quadrilateral of stars to its southeast. Continuation of a line from Vega through Altair indicates the zodiacal constellation of **Capricornus**.

The Autumn Constellations

During the autumn season, the most striking feature is the "Great Square of Pegasus", an almost perfect rectangle on the sky, forming the main body of the constellation of **Pegasus**. However, the star at the northeastern corner, **Alpheratz**, is actually α Andromedae, and part of the adjacent constellation of **Andromeda**. A line from **Scheat** (β Pegasi) at the northeastern corner of the Square, through **Matar** (η Pegasi), points in the general direction of Cygnus. A crooked line of stars leads from **Markab** (α Pegasi) through **Homam** (ζ Pegasi) and **Biham** (θ Pegasi) to **Enif** (ε Pegasi). A line from Markab through the last star in the Square, **Algenib** (γ Pegasi) points in the general direction of the five stars, including **Menkar** (α Ceti) that form the "tail" of the constellation of **Cetus** (the Whale). A ring of seven stars lying below the southern side of the Great Square is known as "the Circlet", part of the constellation of **Pisces** (the Fishes).

Extending the line of the western side of the Great Square towards the south leads to the isolated bright star, **Fomalhaut** (α Piscis Austrini), in the Southern Fish. Following the line of the eastern side of the Great Square towards the north leads to Cassiopeia while, in the other direction, it points towards **Diphda** (β Ceti), which is actually the brightest star in Cetus.

Three bright stars leading northeast from Alpheratz form the main body of the constellation of **Andromeda**. Continuation of that line leads towards the constellation of **Perseus** and **Mirfak** (α Persei). Running southwards from Mirfak is a chain of stars, one of which is the famous variable star **Algol** (β Persei). Farther east, an arc of stars leads to the prominent cluster of the **Pleiades**, in the constellation of **Taurus**.

Between Andromeda and Cetus lie the two small constellations of **Triangulum** (the Triangle) and **Aries** (the Ram).

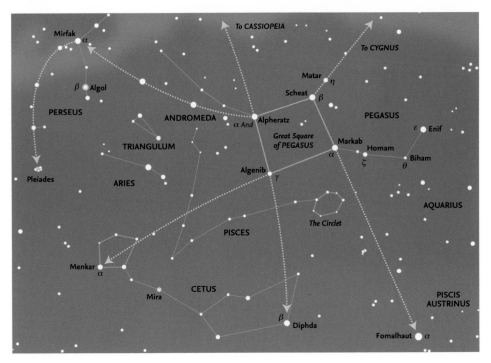

The Moon at First Quarter.

The Moon

The monthly pages include diagrams showing the phase of the Moon for every day of the month, and also indicate the day in the **lunation** (or **age** of the Moon), which begins at New Moon. Although the main features of the surface – the light highlands and the dark maria (seas) – may be seen with the naked eye, far more features may be detected with the use of binoculars or any telescope. The many craters are best seen when they are close to the **terminator** (the boundary between the illuminated and the non-illuminated areas of the surface), when the Sun rises or sets over any particular region of the Moon and the crater walls or central peaks cast strong shadows. Most features become difficult to see at Full Moon, although this is the best time to see the bright ray systems surrounding certain craters. Accompanying the Moon map on the following pages is a list of prominent features, including the days in the lunation when features are normally close to the terminator and thus easiest to see. A few bright features

such as Linné and Proclus, visible when well illuminated, are also listed. One feature, Rupes Recta (the Straight Wall) is readily visible only when it casts a shadow with light from the east, appearing as a light line when illuminated from the opposite direction.

The dates of visibility vary slightly through the effects of **libration**. Because the Moon's orbit is inclined to the Earth's equator and also because it moves in an ellipse, the Moon appears to rock slightly from side to side (and nod up and down). Features near the **limb** (the edge of the Moon) may vary considerably in their location and visibility. (This is easily noticeable with Mare Crisium and the craters Tycho and Plato.) Another effect is that at crescent phases before and after New Moon, the normally non-illuminated portion of the Moon receives a certain amount of light, reflected from the Earth. This **Earthshine** may enable certain bright features (such as Aristarchus, Kepler and Copernicus) to be detected even though they are not illuminated by sunlight.

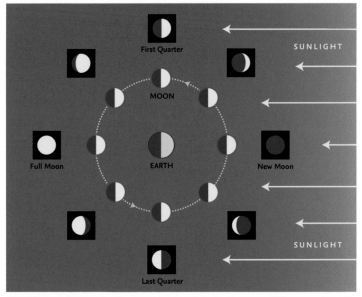

The Moon phases. *During its orbit around the Earth we see different portions of the illuminated side of the Moon's surface.*

Map of the Moon

*The numbers indicate the age of
the Moon when features are
usually best visible.*

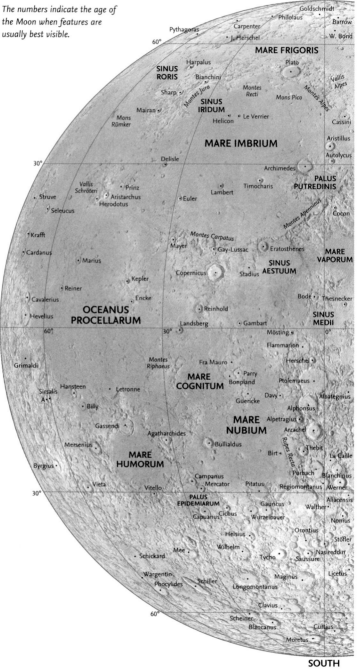

18

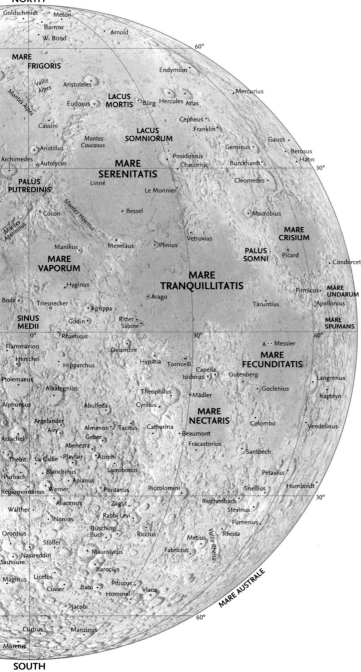

NORTH

Goldschmidt · Meton
Barrow
W. Bond · Arnold
60°
MARE
FRIGORIS · Endymion ·
Vallis · Aristoteles · Mercurius ·
Alpes
Eudoxus · **LACUS** · Bürg · Hercules · Atlas
Cassini · **MORTIS**
Cepheus ·
Aristillus · Montes · **LACUS** · Franklin · Gauss ·
Caucasus · **SOMNIORUM** · Geminus · Berosus ·
Archimedes · Autolycus · Posidonius · Hahn
MARE · Chacornac · Burckhardt · 30°
PALUS · **SERENITATIS** · Cleomedes ·
PUTREDINIS · Linné · Le Monnier ·
Cocon · · Bessel · Macrobius ·
Montes Hoemus · Vetruvius · **MARE**
Manilius, · Menelaus · Plinius · **CRISIUM**
MARE · **PALUS** · Picard
VAPORUM · **SOMNI** · Condorcet
Hyginus · **MARE** · Firmicus · **MARE**
Bode · · Triesnecker · Arago · **TRANQUILLITATIS** · **UNDARUM**
Agrippa · Taruntius · Apollonius
SINUS · Godin · Ritter · **MARE**
MEDII · Sabine · **SPUMANS**
0° · Rhaeticus · 30° · 60° · **E**
Flammarion · Delambre · A · Messier
Herschel · Hypatia · Torricelli · **MARE**
Hipparchus · Capella · **FECUNDITATIS**
Ptolemaeus · Isidorus · Gutenberg · Langrenus
Albategnius · Theophilus · Mädler · Goclenius · Kapteyn
Alphonsus · Abulfeda · Cyrillus · **MARE**
Argelander · Almanon · Tacitus · Catharina · **NECTARIS** · Colombo · Vendelinus
Airy · Geber · Beaumont
Arzachel · Abenezra · Fracastorius
Thebit · La Caille · Playfair · Azophi · Santbech
Purbach · Blanchinus · Sacrobosco · Petavius ·
Apianus · Snellius · Humboldt
Regiomontanus · Werner · Pontanus · Piccolomini
Aliacensis · Zagut · Reichenbach · 30°
Walther · Rabbi Levi · Stevinus ·
Nonius · Furnerius ·
Orontius · Büsching · Metius · Rheita
Stöfler · Buch · Riccius · Fabricius
Nasireddin · Maurolycus
Saussure
Barocius
Maginus · Licetus
Cuvier · Baco · Pitiscus · Vlacq
Hommel
Jacobi
60°
Curtius · Manzinus
Moretus

SOUTH

Macrobius	4:18
Mädler	5:19
Maginus	8:22
Manilius	7:21
Mare Crisium	2-3:16-17
Maurolycus	6:20
Mercator	10:24
Metius	4:18
Meton	6:20
Mons Pico	8:22
Mons Piton	8:22
Mons Rümker	12:26
Montes Alpes	6-8:21
Montes Apenninus	8
Orontius	8:22
Pallas	8:22
Petavius	3:17
Philolaus	9:23
Piccolomini	5:19
Pitatus	8:22
Pitiscus	5:19
Plato	8:22
Plinius	6:20
Posidonius	5:19
Proclus	14:18
Ptolemaeus	8:22
Purbach	8:22
Pythagoras	12:26
Rabbi Levi	6:20
Reinhold	9:23
Rima Ariadaeus	6:20
Rupes Recta	8
Saussure	8:22
Scheiner	10:24
Schickard	12:26
Sinus Iridum	10:24
Snellius	3:17
Stöfler	7:21
Taruntius	4:18
Thebit	8:22
Theophilus	5:19
Timocharis	8:22
Triesnecker	6-7:21
Tycho	8:22
Vallis Alpes	7:21
Vallis Schröteri	11:25
Vlacq	5:19
Walther	7:21
Wargentin	12:27
Werner	7:21
Wilhelm	9:23
Zagut	6:20

Eclipses in 2021

Lunar eclipses

There are two lunar eclipses in 2021, one total and one that is technically partial, although most of the Moon will pass within the *umbra* and appear dark. These eclipses are shown in the diagrams opposite. In both cases, mid-eclipse occurs over the middle of the Pacific Ocean. The first (on May 26) will be visible from the central and southern Pacific, and the second (on November 19) from the northern region. During that partial eclipse, a small portion will remain within the *penumbra*.

Solar eclipses

There are two solar eclipses in 2021, the central lines being visible only from the Earth's polar regions. The **annular** solar eclipse of June 10 occurs over Arctic Canada (with maximum eclipse at 10:43 UT over northwest Greenland), continuing over the Arctic Ocean and ending over Siberia. The partial phases will be visible over the northern Atlantic, northern Europe and the last stages after sunrise in parts of the northeastern United States and Canada.

The second eclipse on December 4 is in the far south. It begins and ends over the Southern Ocean and crosses West Antarctica. Maximum eclipse occurs (at 07:33 UT) over the coast of the Weddell Sea. The partial phases are visible over a large region of the Southern Ocean and East Antarctica.

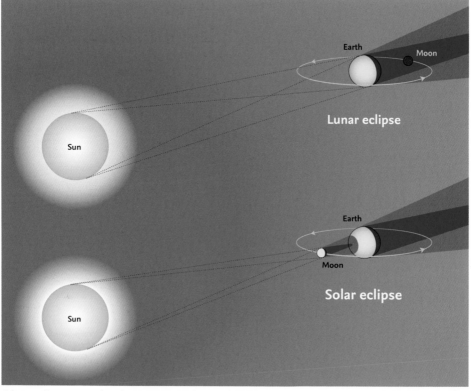

A lunar eclipse occurs when the Moon passes through the Earth's shadow (top). When it passes between the Earth and the Sun (bottom) there is a solar eclipse.

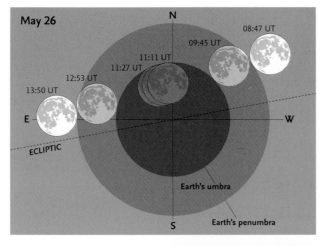

May 26

N

08:47 UT

09:45 UT

11:11 UT
11:27 UT

12:53 UT

13:50 UT

E

W

ECLIPTIC

Earth's umbra

Earth's penumbra

S

A total lunar eclipse occurs on May 26. Generally visible from the Pacific, just the beginning is visible from the West Coast of North America. Most of the eclipse is seen from Australia and New Zealand, including mid-eclipse (at 11:18 UT). The very end becomes visible from Indonesia and eastern Asia.

The lunar eclipse on November 19 is partial, because a very small portion of the Moon remains outside the umbra, in the penumbra. This eclipse will be visible from North America; at mid eclipse (09:03 UT) the Moon will be low for observers on the East Coast. Only the extreme West Coast (and eastern Asia, together with northern Australia) will see the final phases.

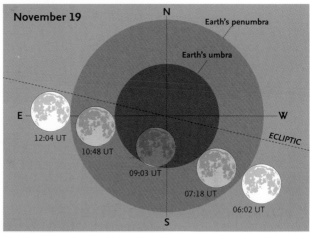

November 19

N

Earth's penumbra

Earth's umbra

E

W

12:04 UT

ECLIPTIC

10:48 UT

09:03 UT

07:18 UT

06:02 UT

S

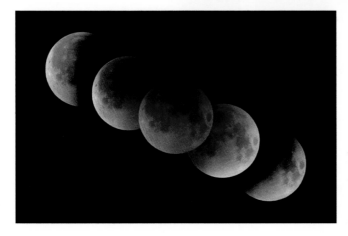

A sequence of five images of the Moon, taken by Akira Fuji, showing the progression of the Moon (from right to left: west to east) through the Earth's shadow.

The Planets in 2021

Mercury and Venus

Although **Venus** may be prominent in the sky for many months at a time, the innermost planet, **Mercury**, is readily visible only at certain elongations. Not every one is favorable. In 2021, for example, Mercury comes to greatest elongation seven times, but it is most visible at five of those only. These occasions are shown on the diagrams below.

Venus is prominent throughout 2021, but has a single greatest elongation on October 29, shown in the last diagram on this page.

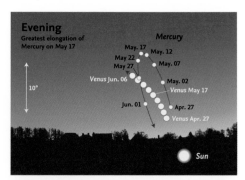

Evening
Greatest elongation of Mercury on May 17

Mercury

May. 17
May 22
May 27
May. 12
May. 07
Venus Jun. 06
May. 02
Venus May 17
Jun. 01
Apr. 27
Venus Apr. 27
10°
Sun

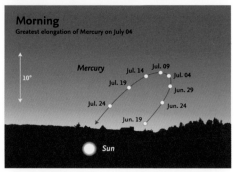

Morning
Greatest elongation of Mercury on July 04

Mercury
Jul. 14
Jul. 09
Jul. 04
Jul. 19
Jun. 29
Jul. 24
Jun. 24
Jun. 19
10°
Sun

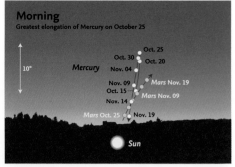

Morning
Greatest elongation of Mercury on October 25

Oct. 25
Oct. 30
Oct. 20
Mercury
Nov. 04
Nov. 09
Mars Nov. 19
Oct. 15
Mars Nov. 09
Nov. 14
Mars Oct. 25
Nov. 19
10°
Sun

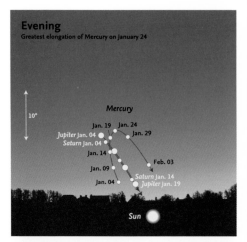

Evening
Greatest elongation of Mercury on January 24

Mercury

Jan. 19 Jan. 24
Jupiter Jan. 04
Jan. 29
Saturn Jan. 04
Jan. 14
Feb. 03
Jan. 09
Saturn Jan. 14
Jan. 04
Jupiter Jan. 19
10°
Sun

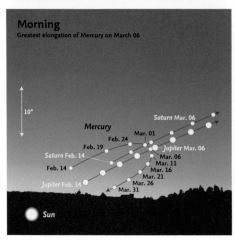

Morning
Greatest elongation of Mercury on March 06

Saturn Mar. 06
Mercury
Mar. 01
Feb. 24
Mar. 06
Feb. 19
Jupiter Mar. 06
Saturn Feb. 14
Mar. 06
Feb. 14
Mar. 11
Mar. 16
Mar. 21
Jupiter Feb. 14
Mar. 26
Mar. 31
10°
Sun

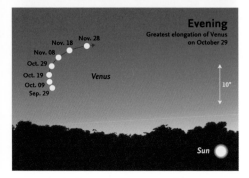

Evening
Greatest elongation of Venus on October 29

Nov. 28
Nov. 18
Nov. 08
Oct. 29
Oct. 19
Venus
Oct. 09
Sep. 29
10°
Sun

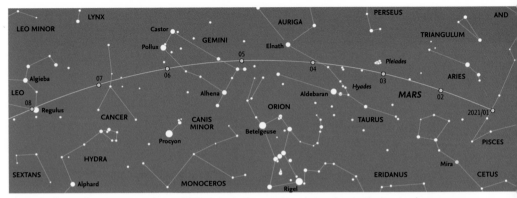

The path of Mars from January to August 2021. Later in the year, Mars is too close to the Sun to observe.

Mars

Because its orbit lies outside that of the Earth, so it takes longer to complete a full orbit, **Mars** does not come to opposition every year, as is the case in 2021. Instead, it remains in the sky for many months at a time, moving slowly along the ecliptic. The planet's path in 2021 is shown on the large chart given here. It begins the year at magnitude -0.2, but then fades to mag. 1.8 in August, when it becomes too close to the Sun to be visible.

Oppositions occur during a period of **retrograde motion**, when the planet appears to move westwards against the pattern of distant stars. As mentioned, there is no opposition in 2021, and the planet reverted to **direct motion** (eastward) on 17 November 2020.

Because of its eccentric orbit, which carries it at very differing distances from the Sun (and Earth), not all oppositions of Mars are equally favorable for observation. The relative positions of Mars and the Earth are shown here. It will be seen that the opposition of 2018 was very close and thus favorable for observation, and that of 2020 was also reasonably good. By comparison, opposition in 2027 will be at a far greater distance, so the planet will appear much smaller.

Between June 22 and June 25, Mars crosses the open star cluster M44 (Praesepe) in Cancer. All stars brighter than magnitude 8.5 are shown.

The oppositions of Mars between 2018 and 2033. As the illustration shows, there is no opposition in 2021.

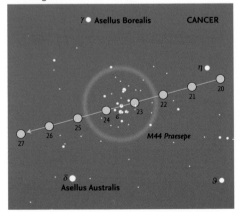

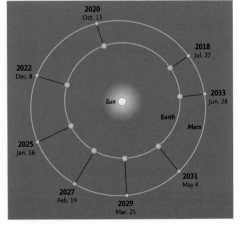

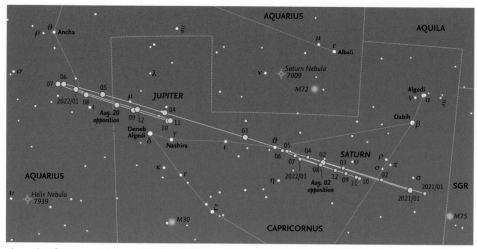

The paths of Jupiter and Saturn in 2021. Background stars are shown down to magnitude 6.5.

Jupiter and Saturn

In 2021, **Jupiter** and **Saturn** are in the same region of sky. Jupiter moves right across **Capricornus** and into **Aquarius**, where it begins to retrograde on June 21. It comes to opposition, just within Aquarius, on August 20, and resumes direct motion on October 30. Saturn, by contrast, remains within Capricornus, retrograding from May 24 to October 19, with opposition on August 2.

Jupiter's four large satellites are readily visible in binoculars. Not all four are visible all the time, sometimes hidden behind the planet or invisible in front of it. **Io**, the closest to Jupiter, orbits in just under 1.8 days, and **Callisto**, the farthest away, takes about 16.7 days. In between are **Europa** (*c*. 3.6 days) and **Ganymede**, the largest, (*c*. 7.1 days). The diagram shows the satellites' motions around the time of opposition.

Jupiter and the four Galilean satellites, photographed by Jan Sandberg with a 10-inch (254-mm) Schmidt-Cassegrain telescope.

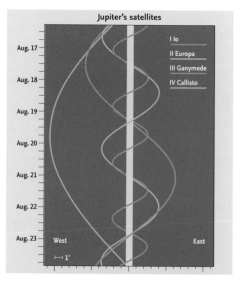

Uranus and Neptune

Uranus begins the year (at mag. 5.7) retrograding in **Aries**, in which constellation it remains for the whole year. It starts direct motion on January 16 and does not begin retrograding until August 20. It reaches opposition (mag. 5.6) on November 4 (at New Moon). It is still retrograding at the end of the year, when it is again mag. 5.7.

Neptune is in **Aquarius** throughout 2021, beginning at mag. 7.9. It moves eastwards until June 27, when it begins to retrograde. It comes to opposition (at mag. 7.8) on September 14 (just after First Quarter, so moonlight should not interfere greatly). It resumes direct motion on December 4, and is mag. 7.9 at the end of the year.

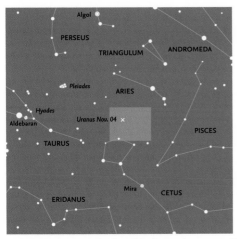

In 2021, Uranus may be found in the southern part of the constellation of Aries.

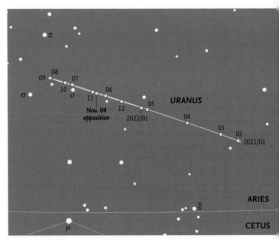

The path of Uranus in 2021. Uranus comes to opposition on November 4. All stars brighter than magnitude 7.5 are shown.

In 2021, Neptune is to be found in the easternmost part of the constellation of Aquarius.

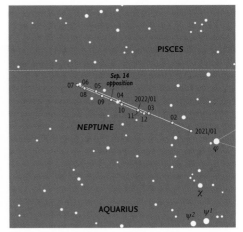

The path of Neptune in 2021. Neptune comes to opposition on September 14. All stars brighter than magnitude 8.5 are shown.

Minor Planets in 2021

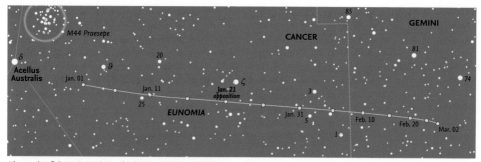

The path of the minor planet (15) Eunomia around its opposition on January 21 (mag. 8.5). Background stars are shown down to magnitude 9.5.

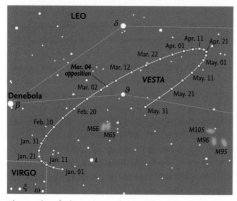

The path of the minor planet (4) Vesta around its opposition on March 4 (mag. 6.0). Background stars are shown down to magnitude 7.5.

In 2021, five minor planets become bright enough to be visible at opposition. These (in date order of opposition) are: *(15) Eunomia* in *Cancer* on January 21; *(4) Vesta* in *Leo* on March 4; *(6) Hebe* in *Aquila* on July 17; *(2) Pallas* in *Pisces* on September 11; and *(1) Ceres*, the dwarf planet on November 27 in *Taurus*. The locations are shown on the small charts at the bottom of page 27. During the oppositions of Eunomia, Vesta and Hebe there will be some interference from moonlight and these minor planets will then be easier to detect some time before or after the opposition.

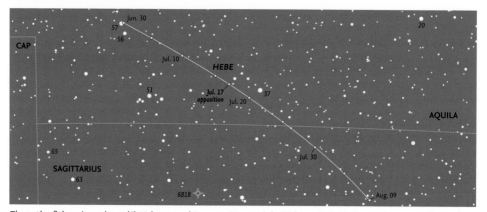

The path of the minor planet (6) Hebe around its opposition on July 17 (mag. 8.4). Background stars are shown down to magnitude 9.5.

The path of the minor planet (2) Pallas around its opposition on September 11 (mag. 8.5). Background stars are shown down to magnitude 9.5. →

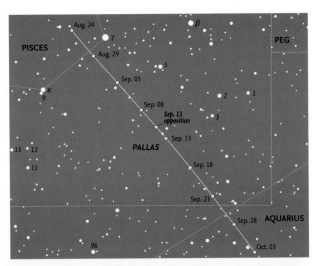

The path of the dwarf planet (1) Ceres around its opposition on November 27 (mag. 7.0). Background stars are shown down to magnitude 8.5.

↓

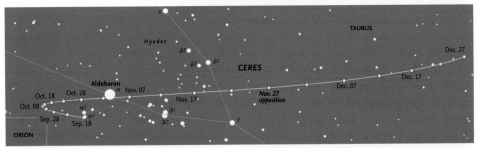

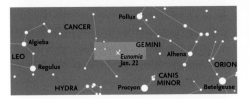

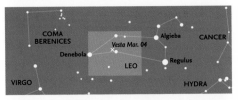

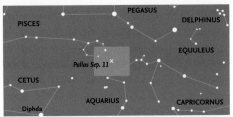

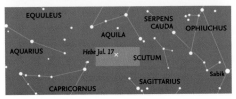

The five small charts show the areas covered by the larger maps on these pages, in a lighter blue. The white cross shows the position of the minor planet at the day of its opposition.

Comets in 2021

Although comets may occasionally become very striking objects in the sky, their occurrence and particularly the existence or length of any tail and their overall magnitude are notoriously difficult to predict. Naturally, it is only possible to predict the return of periodic comets (whose names have the prefix "P"). Many comets appear unexpectedly (these have names with the prefix "C"). Bright, readily visible comets such as C/1995 Y1 Hyakutake, C/1995 01 Hale-Bopp or C/2006 P1 McNaught

(sometimes known as the Great Comet of 2007) are rare. Comet C/2006 P1 McNaught was notable for its multiple tail structure. Most periodic comets are faint and only a very small number ever become bright enough to be readily visible with the naked eye or with binoculars.

The comet most likely to become visible in 2021 is *67P/Churyumov-Gerasimenko*. This is the comet visited and investigated by ESA's Rosetta space probe in 2014–2016.

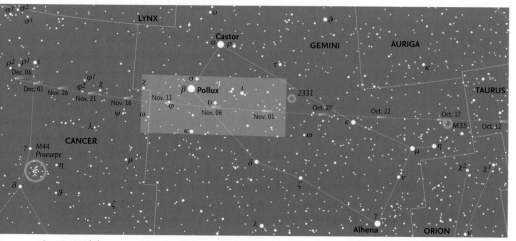

Comet 67P/Churyumov-Gerasimenko may become brighter than its predicted mag. 9.9, around November 6. Its path is shown here from October 12 to December 6. Background stars are shown down to magnitude 8.0. The area covered by the more detailed map below is shown in a lighter blue. On that map, all stars down to magnitude 10.5 are shown.

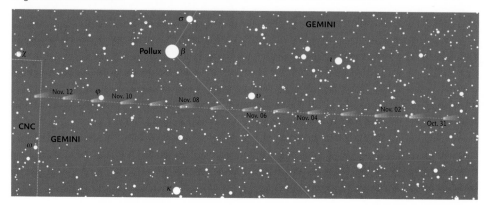

28

Comet 67P/Churyumov-Gerasimenko, photographed by Damian Peach on 21 November 2015, while in the constellation of Leo, approximately one degree east of iota (ι) Leonis.

Comet C/2014 Q2 Lovejoy, which reached naked-eye visibility, photographed on 20 December 2014, when in Columba, by Damian Peach.

Introduction to the Month-by-Month Guide

The monthly charts

The pages devoted to each month contain a pair of charts showing the appearance of the night sky, looking north and looking south. The charts (as with all the charts in this book) are drawn for the latitude of 40°N, so observers farther north will see slightly more of the sky on the northern horizon, and slightly less on the southern. These areas are, of course, those most likely to be affected by poor observing conditions caused by haze, mist or smoke. In addition, stars close to the horizon are always dimmed by atmospheric absorption, so sometimes the faintest stars marked on the charts may not be visible.

The three times shown for each chart require a little explanation. The charts are drawn to show the appearance at 11 p.m. for the 1st of each month. The same appearance will apply an hour earlier (10 p.m.) on the 15th, and yet another hour earlier (9 p.m.) at the end of the month (shown as the 1st of the following month). These times are local time, and apply in all time zones. Where Daylight Saving Time (DST) is used, in 2021 it is introduced on March 14, and ends on November 7. When used, both local time and DST are shown on the charts. Times of specific events are shown in the 24-hour clock of Universal Time (UT) used by astronomers worldwide, and the correct zone time (and DST, where appropriate) may be found from the details inside the front cover.

The charts may be used for earlier or later times during the night. To observe two hours earlier, use the charts for the preceding month; for two hours later, the charts for the next month.

Meteors

Details of specific meteor showers are given in the months when they come to maximum, regardless of whether they begin or end in other months. Note that not all the respective radiants are marked on the charts for that

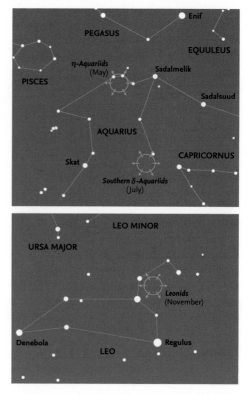

particular month, because the radiants may be below the horizon, or lie in constellations that are not readily visible during the month of maximum. For this reason, special charts for the Eta (η) and Delta (δ) Aquariids (May and July, respectively) and the Leonids (November) are shown above. As explained earlier, however, meteors from such showers may still be seen, because the most effective region for seeing meteors is some 40°–45° away from the radiant, and that area of sky may well be above the horizon. A table of the best meteor showers visible during the year is also given here. The rates given are based on the properties of the meteor streams, and are those that an experienced observer might see under ideal conditions. Generally, the observed rates will be far less.

Shower	Dates of activity 2021	Date of maximum 2021	Possible hourly rate
Quadrantids	December 28 to January 12	January 3–4	120
April Lyrids	April 13–29	April 22	18
η-Aquariids	April 18 to May 27	May 6	40
α-Capricornids	July 2 to August 14	July 30	5
Southern δ-Aquariids	July 13 to August 24	July 30	< 25
Perseids	July 16 to August 23	August 12–13	150
α-Aurigids	August 28 to September 5	September 1	6
Southern Taurids	September 10 to November 20	October 10	< 5
Orionids	October 1 to November 6	October 21	25
Draconids	October 7–11	October 8–9	var.
Northern Taurids	October 20 to December 10	November 12	< 5
Leonids	November 5–29	November 17–18	< 15
Geminids	December 3–16	December 14	120+
Ursids	December 17–26	December 22–23	< 10

Meteors that are brighter than magnitude -4 (approximately the maximum magnitude reached by Venus) are known as *fireballs* or *bolides*. Examples are shown on pages 32, 35 and 77. Fireballs sometimes cause sonic booms that may be heard some time after the meteor is seen.

The photographs

As an aid to identification – especially as some people find it difficult to relate charts to the actual stars they see in the sky – one or more photographs of constellations visible in certain specific months are included. It should be noted, however, that because of the limitations of the photographic and printing processes, and the differences between the sensitivity of different individuals to faint starlight (especially in their ability to detect different colors), and the degree to which they have become adapted to the dark, the apparent brightness of stars in the photographs will not necessarily precisely match that seen by any one observer.

The Moon calendar

The Moon calendar is largely self-explanatory. It shows the phase of the Moon for every day of the month, with the exact times (in Universal Time) of New Moon, First Quarter, Full Moon and Last Quarter. Because the times are calculated from the Moon's actual orbital parameters, some of the times shown

will, naturally, fall during daylight, but any difference is too small to affect the appearance of the Moon on that date. Also shown is the *age* of the Moon (the day in the *lunation*), beginning at New Moon, which may be used to determine the best time for observation of specific lunar features.

The Moon

The section on the Moon includes details of any lunar or solar eclipses that may occur during the month (visible from anywhere on Earth). Similar information is given about any important occultations. Mainly, however, this section summarizes when the Moon passes close to planets or the five prominent stars close to the ecliptic. The dates when the Moon is closest to the Earth (at *perigee*) and farthest from it (at *apogee*) are shown in the monthly calendars, and only mentioned here when they are particularly significant, such as the nearest and farthest during the year.

The planets and minor planets

Brief details are given of the location, movement and brightness of the planets from Mercury to Saturn throughout the month. None of the planets can, of course, be seen when they are close to the Sun, so such periods are generally noted. All of the planets may sometimes lie on the opposite side of the Sun to the Earth (at superior conjunction), but

A fireball (with flares approximately as bright as the Full Moon), photographed against a weak auroral display by D. Buczynski from Tarbat Ness in Scotland on 22 January 2017.

in the case of the inferior planets, Mercury and Venus, they may also pass between the Earth and the Sun (at inferior conjunction) and are normally invisible for a period of time, the length of which varies from conjunction to conjunction. Those two planets are normally easiest to see around either eastern or western elongation, in the evening or morning sky, respectively. Not every elongation is favorable, so although every elongation is listed, only those where observing conditions are favorable are shown in the individual diagrams of events.

The dates at which the superior planets reverse their motion (from direct motion to *retrograde*, and retrograde to direct) and of opposition (when a planet generally reaches its maximum brightness) are given. Some planets, especially distant Saturn, may spend most or all of the year in a single constellation. Jupiter and Saturn are normally easiest to see around opposition, which occurs every year. Mars, by contrast, moves relatively rapidly against the background stars and in some years never comes to opposition.

Uranus is normally magnitude 5.7–5.9, and thus at the limit of naked-eye visibility under exceptionally dark skies, but bright enough to be readily visible in binoculars. Because its orbital period is so long (over 84 years), Uranus moves only slowly along the ecliptic, and often remains within a single constellation for a whole year. The chart on page 25 shows its position during 2021.

Similar considerations apply to Neptune, although it is always fainter (generally magnitude 7.8–8.0), still visible in most binoculars. It takes about 164.8 years to complete one orbit of the Sun. As with Uranus, it frequently spends a complete year in one constellation. Its chart is also on page 25.

In any year, few minor planets ever become bright enough to be detectable in binoculars. Just one, (4) Vesta, on rare occasions brightens sufficiently for it to be visible to the naked eye. Our limit for visibility is magnitude 9.0 and details and charts are given for those objects that exceed that magnitude during the year, normally around opposition. To assist in recognition of a planet or minor planet as it moves against the background stars, the latter are shown to a fainter magnitude than the object at opposition. Minor-planet charts for 2021 are on pages 26 and 27.

The ecliptic charts

Although the ecliptic charts are primarily designed to show the positions and motions of the major planets, they also show the motion of the Sun during the month. The light-tinted area shows the area of the sky that is invisible during daylight, but the darker area gives an indication of which constellations are likely to be visible at some time of the night. The closer a planet is to the border between dark and light, the more difficult it will be to see in the twilight.

The monthly calendar

For each month, a calendar shows details of significant events, including when planets are close to one another in the sky, close to the Moon, or close to any one of five bright stars that are spaced along the ecliptic. The times shown are given in Universal Time (UT), always used by astronomers throughout the year. The tables given inside the front cover may be used to convert UT to that used in any given time zone and (during the summer) to DST.

The diagrams of interesting events

Each month, a number of diagrams show the appearance of the sky when certain events take place. However, the exact positions of celestial objects and their separations greatly depend on the observer's position on Earth. When the Moon is one of the objects involved, because it is relatively close to Earth, there may be very significant changes from one location to another. Close approaches between planets or between a planet and a star are less affected by changes of location, which may thus be ignored.

The diagrams showing the appearance of the sky are drawn for the latitude of 40°N and 90°W, northwest of Springfield, IL, so will be approximately correct for most of North America. However, for an observer farther north (say, Vancouver), a planet or star listed as being north of the Moon will appear even farther north, whereas one south of the Moon will appear closer to it – or may even be hidden (occulted) by it. For an observer at a latitude of less than 40°N, there will be corresponding changes in the opposite direction: for a star or planet south of the Moon, the separation will increase, and for one north of the Moon, the separation will decrease. This is particularly important when occultations occur, which may be visible from one location, but not another. However, there are no major occultations visible in 2021.

Ideally, details should be calculated for each individual observer, but this is obviously impractical. In fact, positions and separations are actually calculated for a theoretical observer located at the center of the Earth.

So the details given regarding the positions of the various bodies should be used as a guide to their location. A similar situation arises with the times that are shown. These are calculated according to certain technical criteria, which need not concern us here. However, they do not necessarily indicate the exact time when two bodies are closest together. Similarly, dates and times are given, even if they fall in daylight, when the objects are likely to be completely invisible. However, such times do give an indication that the objects concerned will be in the same general area of the sky during both the preceding and the following nights.

Data used in this Guide

The data given in this Guide, such as timings and distances between objects, have been computed with a program developed by the US Naval Observatory in Washington D.C., widely regarded as the most accurate computation. As such, the data may differ slightly from information given elsewhere.

Key to the symbols used on the monthy star maps.

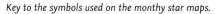

STAR MAGNITUDES

| | -1 | 0 | 1 | 2 | 3 | 4 | Fainter |

Milky Way

OBJECTS

Open Star Cluster   Planetary Nebula   Meteor Radiant

Globular Star Cluster   Bright Nebula   Galaxy

PLANETS

MERCURY
VENUS
MARS
JUPITER
SATURN

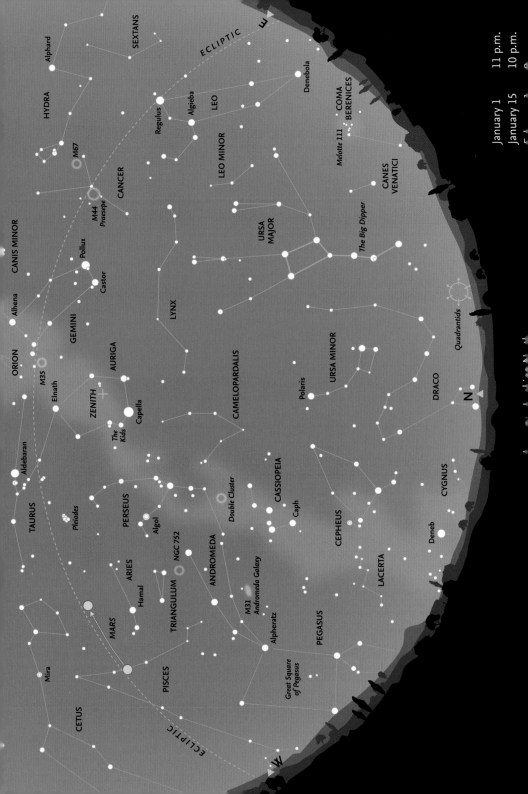

ECLIPTIC

SEXTANS

Alphard

HYDRA

CANIS MINOR

M67

CANCER

M44
Praesepe

Regulus

Algieba

LEO

Denebola

COMA
BERENICES

Melotte 111

LEO MINOR

CANES
VENATICI

The Big Dipper

URSA
MAJOR

Pollux

Castor

GEMINI

LYNX

ORION

Alhena

M35

Elnath

AURIGA

CAMELOPARDALIS

URSA MINOR

Polaris

DRACO

Quadrantids

N

ZENITH

Capella

The
Kids

CASSIOPEIA

Caph

CEPHEUS

CYGNUS

Deneb

Aldebaran

TAURUS

Pleiades

PERSEUS

Algol

Double Cluster

LACERTA

NGC 752

ANDROMEDA

M31
Andromeda Galaxy

Alpheratz

PEGASUS

Great Square
of Pegasus

ARIES

Hamal

TRIANGULUM

MARS

Mira

PISCES

CETUS

ECLIPTIC

W

January – Looking North

Most of the important circumpolar constellations are easy to see in the northern sky at this time of year. *Ursa Major* stands more-or-less vertically above the horizon in the northeast, with the zodiacal constellation of *Leo* rising in the east. To the north, the stars of *Ursa Minor* lie below *Polaris* (the Pole Star). The head of *Draco* is low on the northern horizon, but may be difficult to see unless observing conditions are good. Both *Cepheus* and *Cassiopeia* are readily visible in the northwest, and even the faint constellation of *Camelopardalis* is high enough in the sky for it to be easily visible.

Near the zenith is the constellation of *Auriga* (the Charioteer), with brilliant *Capella* (α Aurigae), directly overhead. Slightly to the west of Capella lies a small triangle of fainter stars, known as "The Kids." (Ancient mythological representations of Auriga show him carrying two young goats.) Together with the northernmost bright star in Taurus, *Elnath* (β Tauri), the body of Auriga forms a large pentagon on the sky, with The Kids lying on the western side. Farther down toward the west are the constellations of *Perseus* and *Andromeda*, and the Great Square of *Pegasus* is approaching the horizon.

Meteors

One of the strongest and most consistent meteor showers of the year occurs in January: the *Quadrantids*, which are visible December 28 to January 12, with maximum on January 3–4. They are brilliant, bluish and yellowish-white meteors and fireballs, which at maximum may even reach a rate of 120 meteors per hour. At maximum on January 3–4 the Moon is waning gibbous, a few days before Last Quarter, so there will be fairly considerable interference from moonlight. The parent object is minor planet 2003 EH_1.

The shower is named after the former constellation *Quadrans Muralis* (the Mural Quadrant), an early form of astronomical instrument. The Quadrantid meteor radiant, marked on the chart, is now within the northernmost part of *Boötes*, roughly halfway between θ Boötis and τ Herculis. This is very low on the northern horizon, roughly halfway between *Alioth*, the last star in the "handle" of the *Big Dipper* and the head of *Draco*.

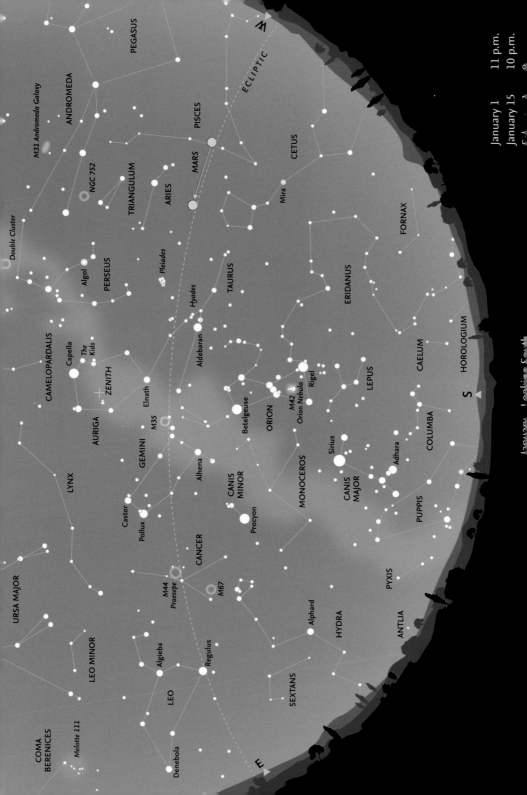

PEGASUS

ANDROMEDA

M31 Andromeda Galaxy

NGC 752

Double Cluster

PERSEUS

Algol

TRIANGULUM

ARIES

PISCES

MARS

CETUS

Mira

FORNAX

Pleiades

TAURUS

Hyades

Aldebaran

Elnath

ERIDANUS

CAMELOPARDALIS

Capella

The Kids

ZENITH

AURIGA

M35

ORION

Betelgeuse

M42
Orion Nebula

Rigel

LEPUS

CAELUM

HOROLOGIUM

S

GEMINI

Castor

Pollux

Alhena

CANIS MINOR

Procyon

MONOCEROS

Sirius

CANIS MAJOR

Adhara

COLUMBA

PUPPIS

LYNX

CANCER

M44
Praesepe

M67

PYXIS

ANTLIA

HYDRA

Alphard

URSA MAJOR

LEO MINOR

Algieba

LEO

Regulus

SEXTANS

COMA BERENICES

Melotte 111

Denebola

E

W

ECLIPTIC

January 1   11 p.m.
January 15   10 p.m.

January — Looking South

January – Looking South

The southern sky is dominated by **Orion**, prominent during the winter months, visible at some time during the night. It is highly distinctive, with a line of three stars that form the "Belt". To most observers, the bright star **Betelgeuse** (α Orionis), shows a reddish tinge, in contrast to the brilliant bluish-white **Rigel** (β Orionis). The three stars of the belt lie directly south of the celestial equator. A vertical line of three "stars" forms the "Sword" that hangs south of the Belt. With good viewing, the central "star" appears as a hazy spot, even to the naked eye, and is actually the **Orion Nebula**. Binoculars reveal the four stars of the Trapezium, which illuminate the nebula

Orion's Belt points up to the northwest towards **Taurus** and orange-tinted **Aldebaran** (α Tauri). Close to Aldebaran, there is a conspicuous "V" of stars, called the **Hyades** cluster. (Despite appearances, Aldebaran is not part of the cluster.) Farther along, the same line from Orion passes below a bright cluster of stars, the **Pleiades**, or Seven Sisters. Even the smallest pair of binoculars reveals this as a beautiful group of bluish-white stars. The two most conspicuous of the other stars in Taurus lie directly above Orion, and form an elongated triangle with Aldebaran. The northernmost, **Elnath** (α Tauri), was once considered to be part of the constellation of **Auriga**.

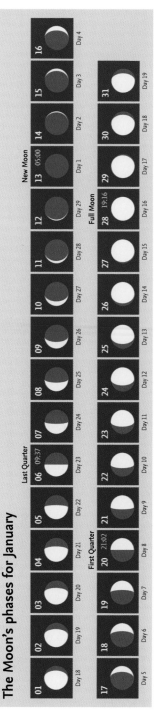

The constellation of Orion dominates the sky during this period of the year, and is a useful starting point for recognizing other constellations in the southern sky. Here, orange Betelgeuse, blue-white Rigel and the pinkish Orion Nebula are prominent. Orion can be found in the southern part of the sky (see page 36).

The Moon's phases for January

Last Quarter

| 01 | 02 | 03 | 04 | 05 | 06 09:37 | 07 |
| Day 18 | Day 19 | Day 20 | Day 21 | Day 22 | Day 23 | Day 24 |

First Quarter

| 08 | 09 | 10 | 11 | 12 | 13 05:00 | 14 |
| Day 25 | Day 26 | Day 27 | Day 28 | Day 29 | Day 1 | Day 2 |

New Moon

| 15 | 16 | 17 | 18 | 19 | 20 21:02 | 21 |
| Day 3 | Day 4 | Day 5 | Day 6 | Day 7 | Day 8 | Day 9 |

Full Moon

| 22 | 23 | 24 | 25 | 26 | 27 | 28 19:16 | 29 |
| Day 10 | Day 11 | Day 12 | Day 13 | Day 14 | Day 15 | Day 16 | Day 17 |

| 30 | 31 |
| Day 18 | Day 19 |

January – Moon and Planets

The Earth

The Earth reaches perihelion (the closest point to the Sun in its annual orbit) on 2 January 2021, at 13:51 Universal Time. Its distance is then 0.983257 AU (147,093,163 km).

The Moon

On January 2, the waning gibbous Moon passes 4.7° north of **Regulus** in **Leo**. On January 6, at Last Quarter, it is north of **Spica** in **Virgo**. On January 10, just before New Moon, it passes 6.0° north of **Antares**. On January 21 it is 5.1° south of **Mars** and then 3.3° south of **Uranus**. On January 24 (waxing gibbous) it passes 4.8° north of **Aldebaran** in **Taurus**, and then, on January 27 (one day before Full) it is south of **Pollux** in **Gemini**. On January 30, waning gibbous, it is again north of Regulus.

The planets

In the evening sky, **Mercury** reaches greatest eastern elongation at mag. -0.7 on January 24. **Venus** (mag. -3.9), initially in **Ophiuchus**, moves closer to the Sun. **Mars** moves from **Pisces** to **Aries**, fading from mag. -0.2 to 0.4. **Jupiter** (mag. -2.0) is in **Capricornus**, too close to the Sun to be visible. **Saturn**, also in Capricornus, is invisible near the Sun. **Uranus** (mag. 5.7 to 5.8) is initially retrograding in Aries, but resumes direct motion on January 16. **Neptune** (mag. 7.9) is in **Aquarius**, where it remains for the whole year. On January 21, minor planet **(15) Eunomia** is at opposition (mag. 8.5) in **Cancer** (see the chart on page 26).

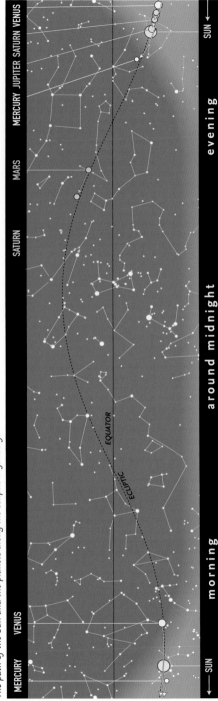

The path of the Sun and the planets along the ecliptic in January.

Date	Time	Event
1–12		Quadrantid meteor shower
2	13:51	Earth at perihelion (0.983257 AU = 147,093,163 km)
2	22:28	Regulus 4.7°S of Moon
3–04		Quadrantid shower maximum
6	09:37	Last Quarter
6	18:33	Spica 7.0°S of Moon
9	15:37	Moon at perigee (367,387 km)
9	21:00 *	Mercury 1.7°S of Saturn
10	02:38	Antares 6.0°S of Moon
11	11:00 *	Mercury 1.5°S of Jupiter
11	20:09	Venus 1.5°N of Moon
13	05:00	New Moon
13	20:52	Saturn 3.2°N of Moon
14	01:27	Jupiter 3.3°N of Moon
14	08:13	Mercury 2.3°N of Moon
17	06:14	Neptune 4.5°N of Moon
20	21:02	First Quarter
20	05:37	Mars 5.1°N of Moon
21	06:24	Uranus 3.3°N of Moon
21	13:11	Moon at apogee (404,360 km)
21	18:37	Minor planet (15) Eunomia at opposition (mag. 8.5)
22	00:00 *	Uranus 1.7°S of Mars
24	01:57	Mercury at greatest elongation (18.6°E, mag. -0.7)
24	03:01	Saturn in conjunction with the Sun
24	05:13	Aldebaran 4.8°S of Moon
27	16:18	Pollux 3.8°N of Moon
28	19:16	Full Moon
29	01:40	Jupiter in conjunction with the Sun
30	05:25	Regulus 4.6°S of Moon

These objects are close together for an extended period at this time.

Evening 10:30 p.m.

January 2 • The Moon in the company of Regulus and Algieba. Denebola is close to the horizon.

Morning 6:45 a.m.

January 9–11 • Shortly before sunrise, the Moon passes Antares, Sabik and Venus (mag. -3.9), low in the southeast.

Evening 5:20 p.m.

January 14 • The narrow crescent Moon with Mercury, Jupiter and Saturn shortly after sunset.

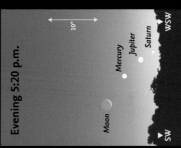

Evening 11 p.m.

January 20 • The Moon is close to Mars and Uranus (mag. 5.8). Background stars are shown down to mag. 5.5.

Evening 6 p.m.

January 26–27 • The almost Full Moon passes Castor and Pollux in the eastern sky

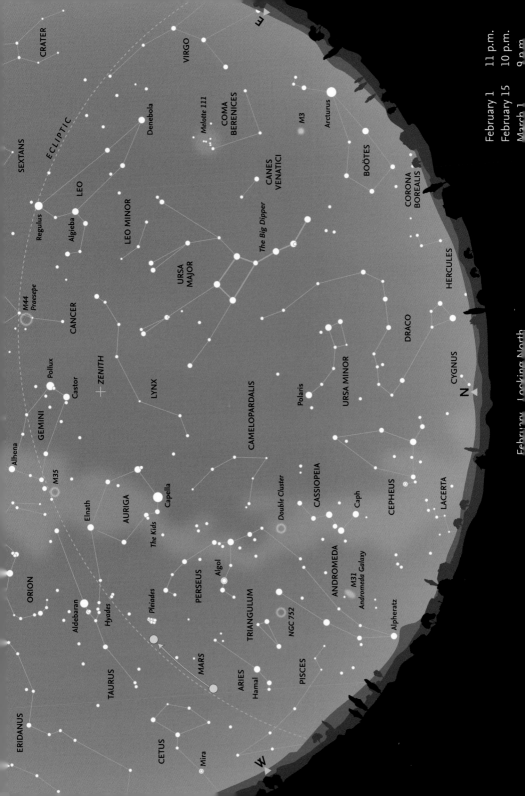

February 1   11 p.m.
February 15   10 p.m.
March 1

February, Looking North

February – Looking North

The months of January and February are probably the best time for seeing the section of the Milky Way that runs in the northern and western sky from **Cygnus**, low on the northern horizon, through **Cassiopeia**, **Perseus** and **Auriga** and then down through **Gemini** and **Orion**. Although not as readily visible as the denser star clouds of the summer Milky Way, on a clear night so many stars may be seen that even a distinctive constellation such as **Cassiopeia** is not immediately obvious.

The head of **Draco** is now higher in the sky and easier to recognize. **Deneb** (α Cygni), the brightest star in **Cygnus**, may just be visible almost due north at midnight, early in the month, if the sky is very clear and the horizon clear of obstacles. **Vega** (α Lyrae) in **Lyra** is so low that it is difficult to see, but may become visible later in the night. The constellation of **Boötes** – sometimes described as shaped like a kite, an ice-cream cone, or the letter "p" – with orange-tinted **Arcturus** (α Boötis), is beginning to clear the eastern horizon. Arcturus, at magnitude -0.05, is the brightest star in the northern hemisphere of the sky. The inconspicuous constellation of **Coma Berenices** is now well above the horizon in the east. The concentration of faint stars at the northwestern corner somewhat resembles a tiny, detached portion of the Milky Way. This is Melotte 111, an open star cluster (which is sometimes called the Coma Cluster, but must not be confused with the important Coma Cluster of galaxies, Abell 1656, mentioned on page 55).

On the other side of the sky, in the northwest, most of the constellation of **Andromeda** is still easily seen, although **Alpheratz** (α Andromedae), the star that forms the northeastern corner of the Great Square of Pegasus – even though it is actually part of Andromeda – is becoming close to the horizon and more difficult to detect. High overhead, at the zenith, try to make out the very faint constellation of **Lynx**. It was introduced in 1687 by the famous astronomer Johannes Hevelius to fill the largely blank area between **Auriga**, **Gemini** and **Ursa Major**, and is reputed to be so named because one needed the eyes of a lynx to detect it.

A very large, and frequently ignored, open star cluster, Melotte 111, also known as the Coma Cluster, is readily visible in the eastern sky during February.

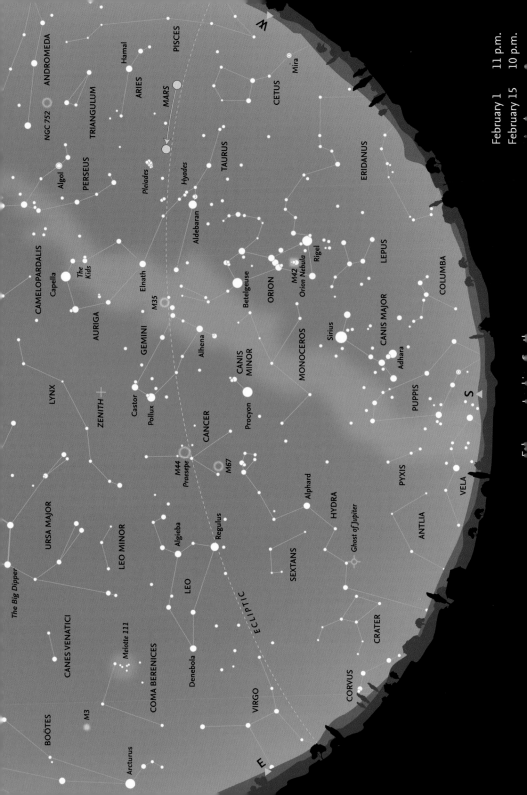

February 1   11 p.m.
February 15   10 p.m.

February – Looking South

Apart from Orion, the most prominent constellation visible this month is **Gemini**, with its two lines of stars running southwest towards **Orion**. Many people have difficulty in remembering which is which of the two stars **Castor** and **Pollux**. Castor (α Geminorum), the fainter star (mag. 1.9), is closer to the North Celestial Pole. Pollux (β Geminorum) is the brighter of the two (mag. 1.2), but is farther away from the Pole. Pollux is one of the first-magnitude stars that may be occulted by the Moon. Castor is remarkable because it is actually a multiple system, consisting of no less than six individual stars.

Orion's belt points down to the southeast towards **Sirius**, the brightest star in the sky (at magnitude -1.4) in the constellation of **Canis Major**, the whole of which is now clear of the southern horizon. Forming an equilateral triangle with **Betelgeuse** in **Orion** and Sirius in Canis Major is **Procyon**, the brightest star in the small constellation of **Canis Minor**. Between Canis Major and Canis Minor is the faint constellation of **Monoceros**, which actually straddles the Milky Way, which, although faint, has many clusters in this area. Directly east of Procyon is the highly distinctive asterism of six stars that form the

The constellation of Gemini. The two brightest stars are Castor and Pollux, visible on the left-hand side of the photograph.

"head" of **Hydra**, the largest of all 88 constellations, and which trails such a long way across the sky that it is only in mid-March around midnight that the whole constellation becomes visible.

The Moon's phases for February

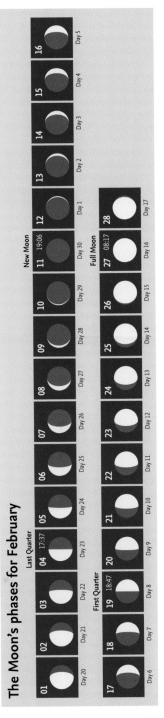

01 — Day 20	02 — Day 21	03 — Day 22
Last Quarter 04 17:37 — Day 23	05 — Day 24	06 — Day 25
07 — Day 26	08 — Day 27	09 — Day 28
10 — Day 29	New Moon 11 19:06 — Day 30	12 — Day 1
13 — Day 2	14 — Day 3	15 — Day 4
16 — Day 5	17 — Day 6	First Quarter 18 — Day 7
19 18:47 — Day 8	20 — Day 9	21 — Day 10
22 — Day 11	23 — Day 12	24 — Day 13
25 — Day 14	26 — Day 15	Full Moon 27 08:17 — Day 16
28 — Day 17		

February – Moon and Planets

The Moon

On February 3, just after midnight one day before Last Quarter, the Moon passes 6.8° north of **Spica** in **Virgo**. On February 6 it is 5.4° north of **Antares** in **Scorpius**. On February 18 it passes 3.7° south of **Mars**. Two days later, it passes south of the **Pleiades** and then 5.0° north of **Aldebaran** in **Taurus**. On January 24 it is 3.7°S of **Pollux**, and on February 26, 4.6°N of **Regulus** in **Leo**.

Occultations

Of the bright stars near the ecliptic that may be occulted by the Moon (**Aldebaran**, **Antares**, **Pollux**, **Regulus** and **Spica**), none are occulted in 2021. There is one occultation of **Mars**, on April 17, partly visible from Southeast Asia.

The planets

Mercury passes inferior conjunction on February 8. **Venus** is invisible in the morning twilight. **Mars**, fading from mag. 0.4 to 0.9, moves eastwards from **Aries** into **Taurus**. **Jupiter** remains too close to the Sun to be visible. **Saturn** is also lost in morning twilight. **Uranus** (mag. 5.8) is in **Aries** and **Neptune** (mag. 7.9 to 8.0) remains in **Aquarius**.

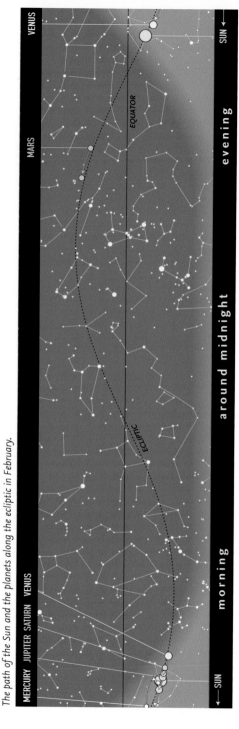

The path of the Sun and the planets along the ecliptic in February.

Calendar for February

03	00:02	Spica 6.8°S of Moon
03	19:03	Moon at perigee (370,127 km)
04	17:37	Last Quarter
06	05:00 *	Saturn 0.4°N of Venus
06	09:04	Antares 5.4°S of Moon
08	13:48	Mercury at inferior conjunction
10	11:10	Saturn 3.4°N of Moon
10	20:25	Venus 3.2°N of Moon
10	21:35	Jupiter 3.7°N of Moon
11	03:17	Mercury 8.3°N of Moon
11	12:00 *	Jupiter 0.4°N of Venus
11	19:06	New Moon
12	17:00 *	Mercury 4.8°N of Venus
13	16:57	Neptune 4.3°N of Moon
13	19:00 *	Mercury 4.2°N of Jupiter
17	15:48	Uranus 3.0°N of Moon
18	10:22	Moon at apogee (404,467 km)
18	22:46	Mars 3.7°N of Moon
19	18:47	First Quarter
20	13:50	Aldebaran 5.0°S of Moon
24	01:42	Pollux 3.7°N of Moon
26	14:33	Regulus 4.6°S of Moon
27	08:17	Full Moon

** These objects are close together for an extended period around this time.*

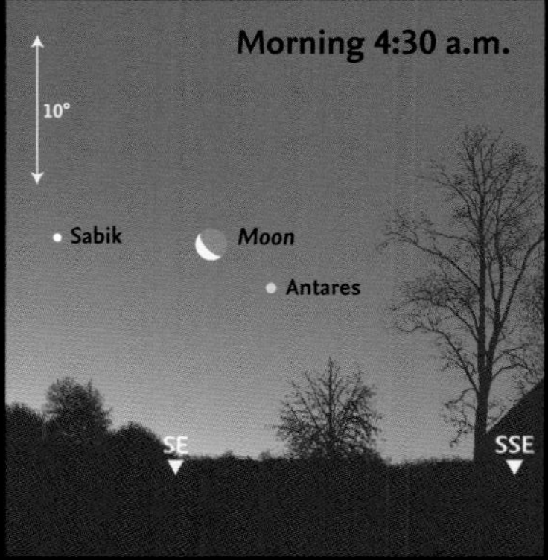

February 6 • *In the early morning, the waning crescent Moon is close to Antares, with Sabik (η Oph) farther east.*

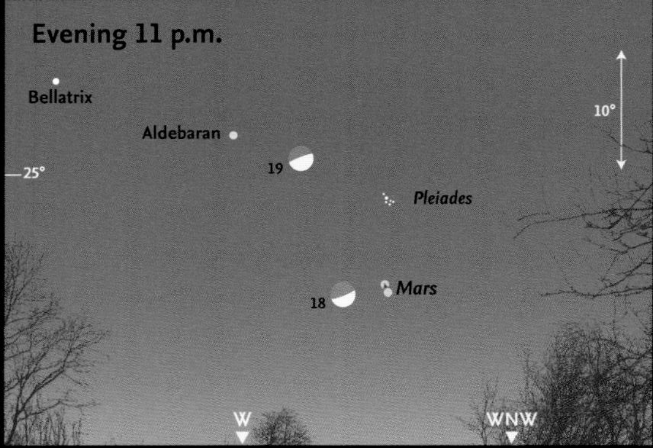

February 18–19 • *The First Quarter Moon passes Mars, the Pleiades and Aldebaran in the west. Mars (mag. 0.7) is just slightly brighter than Aldebaran (mag. 0.9).*

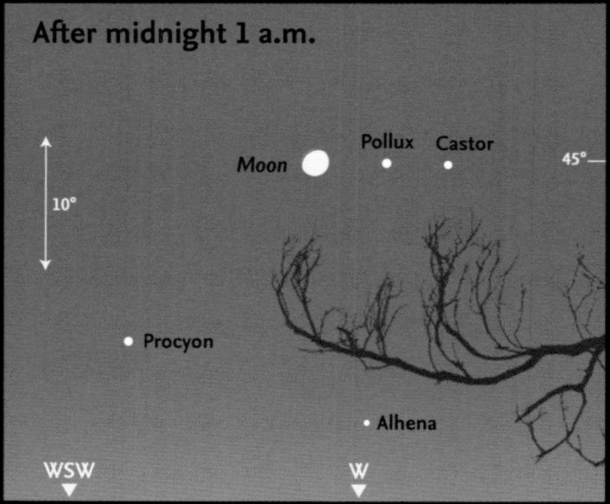

February 24 • *The Moon lines up with Castor and Pollux in the western sky, in the early morning. Procyon and Alhena are nearby.*

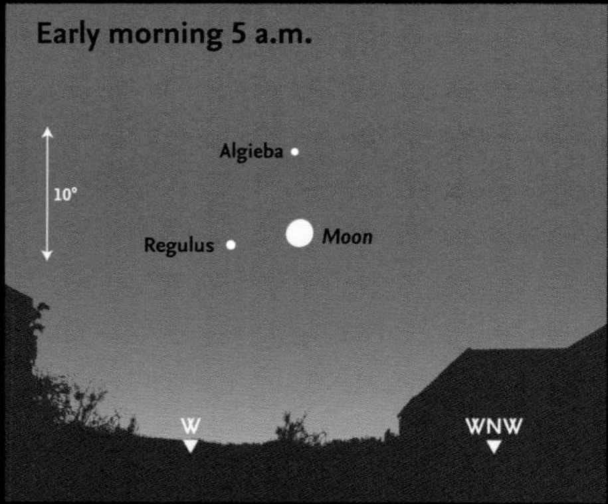

February 26 • *Early in the morning, the almost Full Moon is in the western sky, together with Regulus and Algieba.*

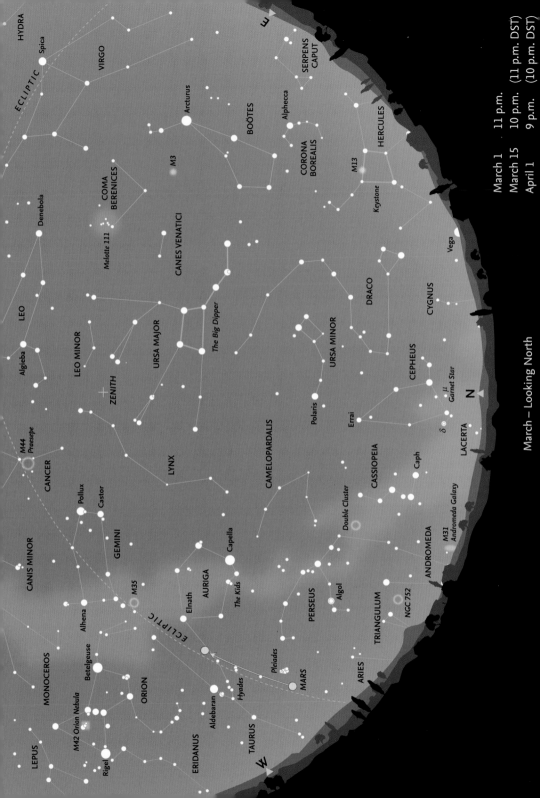

March – Looking North

March 1 . 11 p.m.
March 15 . 10 p.m. (11 p.m. DST)
April 1 . 9 p.m. (10 p.m. DST)

March – Looking North

In March, the Sun crosses the celestial equator on Saturday, March 20, at the vernal equinox, when day and night are of almost exactly equal length, and the northern season of spring is considered to begin. (The hours of daylight and darkness change most rapidly around the equinoxes, in March and September.) It is also in March that Daylight Saving Time (DST) begins in North America (on Sunday, March 14) so the charts show the appearance at 11 p.m. on March 1 and 10 p.m. on April 1. In Europe, Summer Time is introduced two weeks later on March 28.

Early in the month, the constellation of **Cepheus** lies almost due north, with the distinctive "W" of **Cassiopeia** to its west. Cepheus lies across the border of the Milky Way and is often described as like the gable-end of a house and is best seen later in the night or in the month. Despite the large number of stars revealed at the base of the constellation by binoculars, one star stands out because of its deep red color. This is **μ Cephei**, also known as the Garnet Star, because of its striking color. It is a truly gigantic star, a red supergiant, and one of the largest stars known. It is about 2400 times the diameter of the Sun, and if placed in the Solar System would extend beyond the orbit of Saturn. (Betelgeuse, in Orion, is also a red supergiant, but it is "only" about 500 times the diameter of the Sun.)

Another famous, and very important star in Cepheus is **δ Cephei**, which is the prototype for the class of variable stars known as Cepheids. These giant stars show a regular variation in their luminosity, and there is a direct relationship between the period of the changes in magnitude and the stars' actual luminosity. From a knowledge of the period of any Cepheid, its actual luminosity – known as its absolute magnitude – may be derived. A comparison of its apparent magnitude on the sky and its absolute magnitude enables the star's exact distance to be

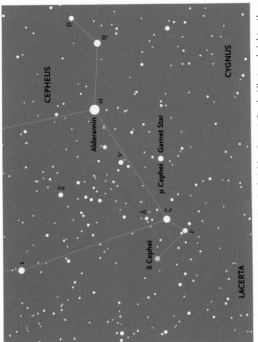

A finder chart for δ Cephei and μ Cephei (the Garnet Star). All stars brighter than magnitude 7.5 are shown.

determined. Once the distances to the first Cepheid variables had been established, examples in more distant galaxies provided information about the scale of the universe. Cepheid variables are the first "rung" in the cosmic distance ladder. Both important stars are shown on the accompanying chart and in the photograph on page 49.

Below Cepheus to the east (to the right), late in the night, it may be possible to catch a glimpse of **Deneb** (α Cygni), just above the horizon. Slightly farther round toward the northeast, **Vega** (α Lyrae) is marginally higher in the sky. From most of the United States, Deneb is just far enough north to be seen at some time during the night (although difficult to see in January and February because it is so low). Vega, by contrast, farther south, is completely hidden during the depths of winter.

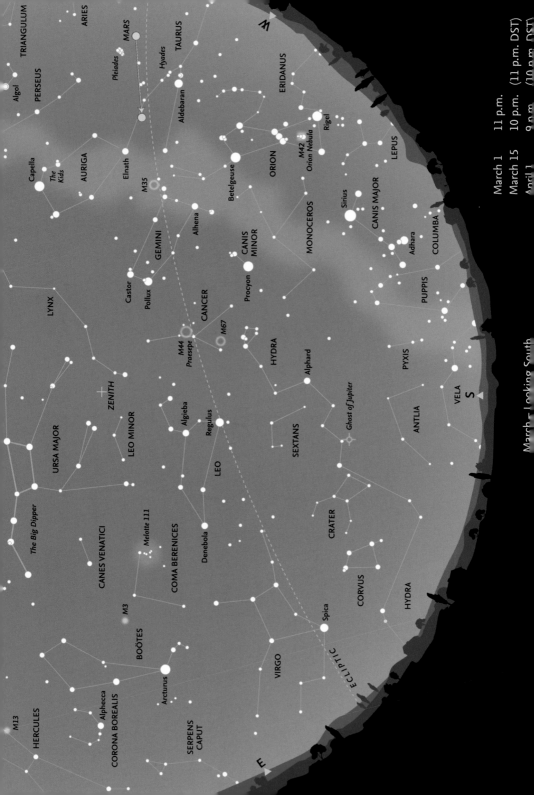

TRIANGULUM
ARIES
MARS
Pleiades
Hyades
TAURUS
PERSEUS
Algol
Aldebaran
ERIDANUS
Elnath
AURIGA
Capella
The Kids
M35
Rigel
M42 Orion Nebula
ORION
LEPUS
Betelgeuse
Alhena
GEMINI
CANIS MAJOR
Castor
Pollux
CANIS MINOR
MONOCEROS
Sirius
Adhara
COLUMBA
CANCER
Procyon
LYNX
M44 Praesepe
M67
PUPPIS
HYDRA
ZENITH
Alphard
PYXIS
LEO MINOR
URSA MAJOR
Algieba
Regulus
SEXTANS
Ghost of Jupiter
VELA
S
ANTLIA
LEO
The Big Dipper
CANES VENATICI
Melotte 111
COMA BERENICES
Denebola
CRATER
M3
HERCULES
BOÖTES
CORONA BOREALIS
Alphecca
Arcturus
SERPENS CAPUT
VIRGO
CORVUS
HYDRA
Spica
M13
ECLIPTIC
E

March – Looking South

March 1   11 p.m.
March 15   10 p.m.   (11 p.m. DST)
April 1   9 p.m.   (10 p.m. DST)

March – Looking South

Due south at 10 p.m. at the beginning of the month, lying between the constellations of *Gemini* in the west and *Leo* in the east, and fairly high in the sky above the head of Hydra, is the faint, and rather undistinguished zodiacal constellation of *Cancer*. Rather like an upside-down letter "Y," it has three "legs" radiating from the center, where there is an open cluster, M44 or *Praesepe* ("the Manger," but also known as "the Beehive"). On a clear night this cluster, known since antiquity, is just a hazy spot to the naked eye, but appears in binoculars as a group of dozens of individual stars.

Also prominent in March is the constellation of *Leo*, with the "backward question mark" (or "Sickle") of bright stars forming the head of the mythological lion. *Regulus* (α Leonis) – the "dot" of the "question mark" or the handle of the sickle and is one of the few first-magnitude stars that may be occulted by the Moon. However, there are no occultations of Regulus in 2021, nor of any of the other four bright stars near the ecliptic.

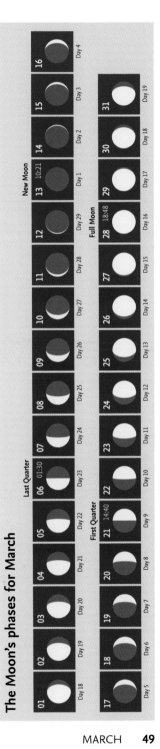

The constellation of Cepheus. The locations of both δ and μ Cephei are shown on the chart on page 47. The bright star near the left edge of the image is Caph (β Cas).

The Moon's phases for March

01 Day 18	02 Day 19	03 Day 20	04 Day 21	05 Day 22	06 01:30 Day 23 Last Quarter	07 Day 24
08 Day 25	09 Day 26	10 Day 27	11 Day 28	12 Day 29	13 10:21 Day 1 New Moon	14 Day 2
15 Day 3	16 Day 4	17 Day 5	18 Day 6	19 Day 7	20 Day 8	21 14:40 Day 9 First Quarter
22 Day 10	23 Day 11	24 Day 12	25 Day 13	26 Day 14	27 Day 15	28 18:48 Day 16 Full Moon
29 Day 17	30 Day 18	31 Day 19				

March – Moon and Planets

The Moon

On March 2, the Moon is 6.6° north of *Spica* in *Virgo*. On March 5, just before Last Quarter, the Moon is 5.2° north of *Antares* in *Scorpius*. On March 19, the Moon passes south of the *Pleiades* and *Mars* and north of *Aldebaran* in *Taurus*. On March 23, the Moon is 3.5° south of *Pollux* in *Gemini*. On March 26 it is 4.7° north of *Regulus* in *Leo*.

The Planets

On March 6, *Mercury* is at greatest western elongation 27.3° away from the Sun, low in morning twilight. *Venus* is also invisible in the twilight as it moves from the morning to the evening sky. It is at superior conjunction on March 26. During the month, *Mars* brightens slightly from mag. 0.9 to 1.3 as it moves eastwards in *Taurus*. *Jupiter* is in the morning twilight in *Capricornus* as is *Saturn*. Uranus (mag. 5.8) is in *Pisces*, close to the Sun. *Neptune* remains in *Aquarius*, also close to the Sun, passing superior conjunction on March 11. Minor planet (4) *Vesta* comes to opposition in *Leo* at mag. 0.6 on March 4, (see chart on page 26).

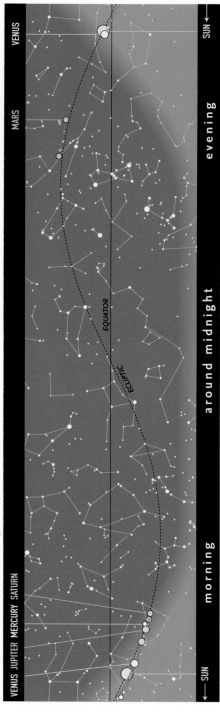

The path of the Sun and the planets along the ecliptic in March.

Calendar for March

02	05:18	Moon at perigee (365,423 km)
02	06:58	Spica 6.6°S of Moon
04	18:08	Minor planet (4) Vesta at opposition (mag. 6.0)
05	07:00 *	Mercury 0.3°N of Jupiter
05	14:29	Antares 5.2°S of Moon
06	01:30	Last Quarter
06	11:22	Mercury at greatest elongation (27.3°W, mag. 0.1)
09	22:57	Saturn 3.7°N of Moon
10	15:36	Jupiter 4.1°N of Moon
11	00:01	Neptune in conjunction with the Sun
11	01:01	Mercury 3.7°N of Moon
13	00:16	Venus 3.9°N of Moon
13	02:38	Neptune 4.3°N of Moon
13	10:21	New Moon
14		North American Daylight Saving Time begins
17	01:50	Uranus 2.7°N of Moon
18	05:03	Moon at apogee (405,253 km)
19	17:47	Mars 1.9°N of Moon
19	21:48	Aldebaran 5.3°S of Moon
20	09:37	Vernal equinox
21	14:40	First Quarter
23	21:00 *	Mars 7.0°N of Aldebaran
23	10:58	Pollux 3.5°N of Moon
26	00:46	Regulus 4.7°S of Moon
26	06:58	Venus at superior conjunction
28		European Daylight Saving Time begins (UK Summer Time)
28	18:48	Full Moon
29	16:22	Spica 6.5°S of Moon
30	06:16	Moon at perigee (360,309 km)

* These objects are close together for an extended period around this time.

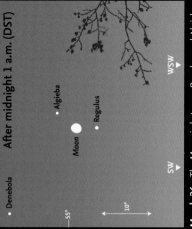

Evening 11 p.m. (DST)

March 19 • The Moon with Mars (mag 1.1) and Aldebaran (mag. 0.9) in the western sky. The Pleiades are lower and farther northwest.

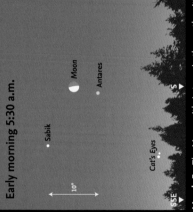

Early morning 5:30 a.m.

March 5 • The Moon with Antares in the south, with Sabik a little farther east. The Cat's Eyes (λ and υ Sco) are below Sabik, closer to the horizon.

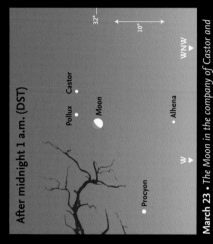

After midnight 1 a.m. (DST)

March 23 • The Moon in the company of Castor and Pollux in the western sky. Alhena (γ Gem) and Procyon in the western sky. Alhena (γ Gem) and Procyon are nearby.

After midnight 1 a.m. (DST)

March 26 • The Moon is between Regulus and Algieba, in the constellation of Leo. Denebola (mag. 2.1) is only slightly fainter than Algieba (mag. 1.9).

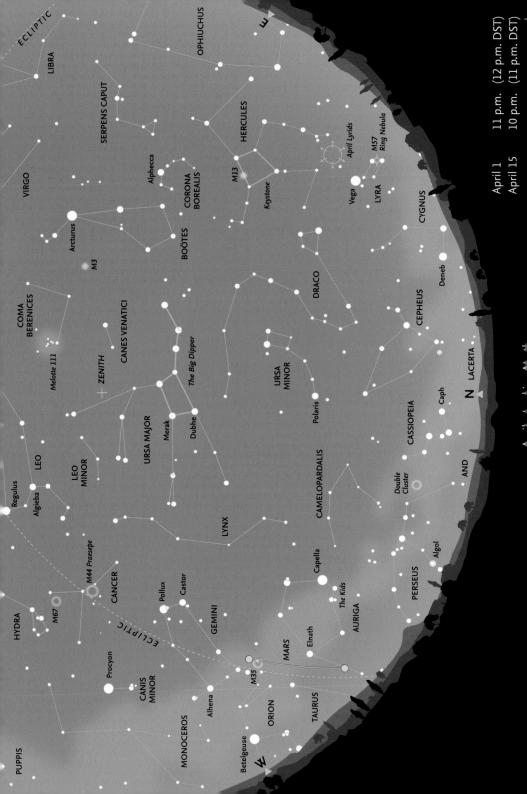

April – Looking North

Cygnus and the brighter regions of the Milky Way are now starting to become visible later in the night. Rising in the northeast is the small constellation of **Lyra** and the distinctive "Keystone" of **Hercules** above it. This asterism is very useful for locating the bright globular cluster M13 (see map on page 59), which lies on one side of the quadrilateral. The winding constellation of **Draco** weaves its way from the four stars that mark its "head," on the border with Hercules, to end at λ Draconis between *Polaris* (α Ursae Minoris) and the "Pointers," *Dubhe* and *Merak* (α and β Ursae Majoris, respectively). **Ursa Major** is "upside down" high overhead, near the zenith. The constellation of **Gemini** stands almost vertically in the west. **Auriga** is still clearly seen in the northwest, but, by the end of the month, the southern portion of *Perseus* is starting to dip below the northern horizon. The very faint constellation of **Camelopardalis** lies in the northwest between Polaris and the constellations of Auriga and Perseus.

Meteors

A moderate meteor shower, the **Lyrids**, peaks on April 22–23. Although the hourly rate is not very high (about 18 meteors per hour), the meteors are fast and some leave persistent trains. This year the maximum occurs when the Moon is waxing gibbous, four days before Full Moon, so there will be considerable interference from moonlight. The parent object is the non-periodic comet C/1861 G1 (Thatcher). Another, stronger shower, the **η-Aquariids**, begins to be active around April 18 (with a waxing crescent Moon) and comes to maximum in early May.

The constellation of Draco, winding round Ursa Minor, with Polaris center left.

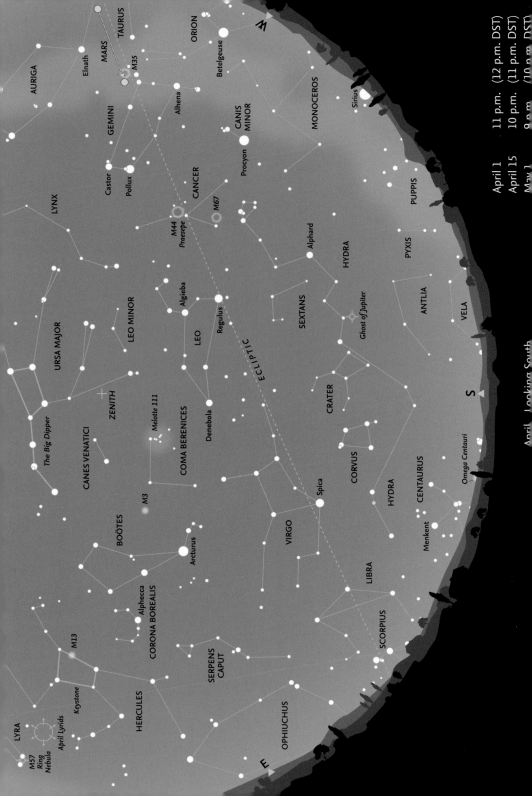

April – Looking South

April 1	11 p.m.	(12 p.m. DST)
April 15	10 p.m.	(11 p.m. DST)
May 1	9 p.m.	(10 p.m. DST)

TAURUS
MARS
Elnath
M35
ORION
Betelgeuse
AURIGA
GEMINI
Alhena
CANIS MINOR
Sirius
MONOCEROS
Castor
Pollux
Procyon
CANCER
LYNX
M44 Praesepe
M67
PUPPIS
Alphard
HYDRA
LEO MINOR
Algieba
LEO
Regulus
SEXTANS
PYXIS
ECLIPTIC
Ghost of Jupiter
ANTLIA
URSA MAJOR
VELA
ZENITH
Melotte 111
Denebola
CRATER
S
The Big Dipper
COMA BERENICES
CANES VENATICI
CORVUS
CENTAURUS
M3
Spica
HYDRA
Omega Centauri
BOÖTES
VIRGO
Menkent
Arcturus
Alphecca
CORONA BOREALIS
LIBRA
M13
Keystone
HERCULES
SERPENS CAPUT
SCORPIUS
April Lyrids
LYRA
M57 Ring Nebula
OPHIUCHUS
E

April – Looking South

Leo is the most prominent constellation in the southern sky in April, and vaguely looks like the creature after which it is named. *Gemini*, with *Castor* and *Pollux*, remains clearly visible in the west, and *Cancer* lies between the two constellations. To the east of Leo, the whole of *Virgo*, with *Spica* (α Virginis) its brightest star, is well clear of the horizon and *Libra* is coming into view. Below Leo and Virgo, the complete length of *Hydra* is visible, running beneath both constellations, with *Alphard* (α Hydrae) halfway between Regulus and the southwestern horizon. Farther east, the two small constellations of *Crater* and the rather brighter *Corvus* lie between Hydra and Virgo. Part of *Centaurus* is now above the horizon.

Boötes and *Arcturus* are prominent in the eastern sky, together with the circlet of *Corona Borealis*, framed by Boötes and the neighboring constellation of *Hercules*. Between Leo and Boötes lies the constellation of *Coma Berenices*, notable for being the location of the open cluster Melotte 111 (see page 41) and the Coma Cluster of galaxies (Abell 1656). There are about 1,000 galaxies in this cluster, which is located near the North Galactic Pole, where we are looking out of the plane of

The distinctive constellation of Leo, with Regulus and "The Sickle" on the west. Algieba (γ Leonis), north of Regulus, appearing double, is a multiple system of four stars.

the Galaxy and are thus able to see deep into space. Only about 10 of the brightest galaxies in the Coma Cluster are visible with the largest amateur telescopes.

The Moon's phases for April

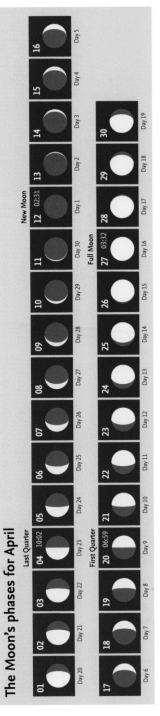

April – Moon and Planets

The Moon

On April 1, the waning gibbous Moon is north of **Antares**. On April 6–7, a waning crescent, it passes south of **Saturn** and **Jupiter** in the morning sky. On April 16–17, the waning crescent Moon passes the **Pleiades**, **Aldebaran** and **Mars**. It passes 0.1° north of Mars on April 17, and there is an occultation, but the end (reappearance) only is visible from southern Asia. On April 19, the Moon (a day before First Quarter), passes 3.2° south of **Pollux** in **Gemini**. The Moon is 6.5° north of **Spica** in **Virgo** on April 26 and 4.8°N of **Antares** in **Scorpius** on April 29.

The planets

Mercury is close to the Sun and comes to superior conjunction on April 19. **Venus** is lost in twilight, it is initially in the morning sky, but moves into the evening later in the month. **Mars** (fading slightly from mag. 1.3 to 1.5) moves from **Taurus** into **Gemini** late in the month. **Jupiter** (mag. -2.1 to -2.2) moves slowly east in the morning sky, entering **Aquarius** at the very end of the month. **Saturn** (also in the morning sky) brightens very slightly from mag. 0.8 to 0.7 and is in **Capricornus**. **Uranus** is in conjunction with the Sun on April 30, in **Aries**, while **Neptune** (mag. 8.0 to 7.9) is in **Aquarius**.

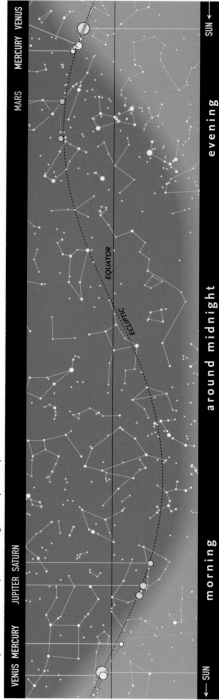

The path of the Sun and the planets along the ecliptic in April.

1	21:19	Antares 4.9°S of Moon
4	10:02	Last Quarter
6	08:29	Saturn 4.0°N of Moon
7	07:17	Jupiter 4.4°N of Moon
9	10:44	Neptune 4.3°N of Moon
1	06:01	Mercury 3.0°N of Moon
2	02:31	New Moon
2	09:48	Venus 2.9°N of Moon
3–29		April Lyrid meteor shower
3	11:41	Uranus 2.5°N of Moon
4	17:46	Moon at apogee (406,119 km)
6	04:41	Aldebaran 5.5°S of Moon
7	12:08	Mars 0.1°N of Moon
8–May27		η-Aquariid meteor shower
9	01:49	Mercury at superior conjunction
9	18:51	Pollux 3.2°N of Moon
0	06:59	First Quarter
2		April Lyrid shower maximum
2	10:18	Regulus 4.9°S of Moon
6	03:13	Spica 6.5°S of Moon
6	09:00 *	Mercury 1.3°N of Venus
6	03:32	Full Moon
7	15:22	Moon at perigee (357,378 km)
9	06:37	Antares 4.8°S of Moon
0	19:54	Uranus at conjunction with the Sun

These objects are close together for an extended

Morning 6 a.m. (DST)

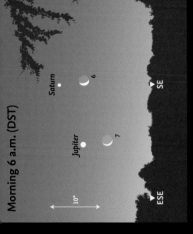

April 6–7 • The waning crescent Moon passes Saturn and Jupiter low in the southeast and shortly before sunrise.

Evening 9 p.m. (DST)

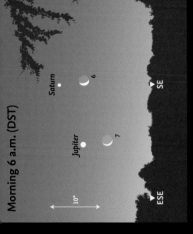

April 19 • High in the western sky, the Moon lines up with Castor and Pollux. Procyon and Alhena are closer

Evening 9 p.m. (DST)

April 15–17 • The waxing crescent Moon passes Aldebaran, Elnath and Mars, high in the west. Mars has now faded to magnitude 1.4.

After midnight 1 a.m. (DST)

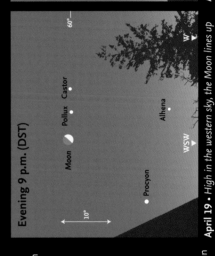

April 29 • After midnight, the Moon is in the southeast in the company of Antares. Sabik is nearby and the

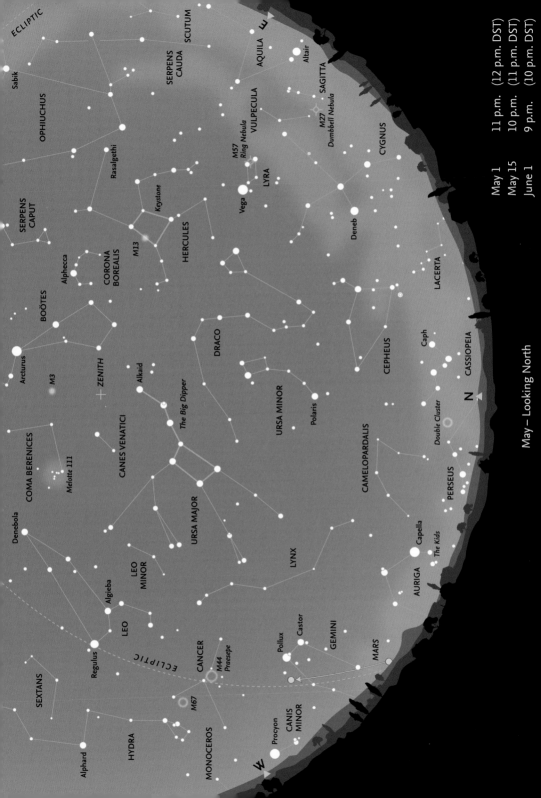

ECLIPTIC

SCUTUM

SERPENS
CAUDA

OPHIUCHUS

AQUILA

Sabik

Altair

SAGITTA

VULPECULA

M27
Dumbbell Nebula

SERPENS
CAPUT

Rasalgethi

M57
Ring Nebula

LYRA

CYGNUS

Vega

Alphecca

Keystone

CORONA
BOREALIS

Deneb

M13

HERCULES

BOÖTES

LACERTA

Arcturus

Caph

M3

DRACO

CEPHEUS

CASSIOPEIA

+ ZENITH

Alkaid

URSA MINOR

CANES VENATICI

The Big Dipper

Polaris

N

CAMELOPARDALIS

Double Cluster

COMA BERENICES

Melotte 111

URSA MAJOR

PERSEUS

Denebola

LEO
MINOR

LYNX

Capella

The Kids

AURIGA

Algieba

LEO

Castor

Pollux

Regulus

GEMINI

ECLIPTIC

MARS

CANCER

M44
Praesepe

SEXTANS

M67

Procyon

CANIS
MINOR

Alphard

HYDRA

MONOCEROS

W

E

May 1   11 p.m. (12 p.m. DST)
May 15   10 p.m. (11 p.m. DST)
June 1   9 p.m. (10 p.m. DST)

May – Looking North

May – Looking North

Cassiopeia is now low over the northern horizon and, to its west, the southern portions of both **Perseus** and **Auriga** have been lost below the horizon, although the **Double Cluster**, between Perseus and Cassiopeia is still clearly visible. The constellations of **Lyra, Cepheus, Ursa Minor** and the whole of **Draco** are well placed in the sky. **Gemini**, with **Castor** and **Pollux**, is sinking towards the western horizon. **Capella** (α Aurigae) and the asterism of **The Kids** are still just clear of the horizon.

In the east, two of the stars of the "Summer Triangle," **Vega** (α Lyrae) and **Deneb** (α Cygni), are clearly visible, and the third star, **Altair** in **Aquila**, is beginning to climb above the horizon. The whole of **Cygnus** is now visible. The sprawling constellation of **Hercules** is high in the east and the brightest globular cluster in the northern hemisphere, M13, is visible to the naked eye on the western side of the asterism known as the **Keystone.**

Three faint constellations may be identified before the lighter nights of summer make them difficult objects. Below Cepheus, low in the northeastern sky is the zig-zag constellation of **Lacerta**, while to the west, above Perseus and Auriga is **Camelopardalis** and, farther west, the line of faint stars forming **Lynx.**

Later in the night (and in the month) the westernmost stars of **Pegasus** begin to come into view, while the stars of **Andromeda** start to appear on the northeastern horizon. High overhead, **Alkaid** (η Ursae Majoris), the last star in the "tail" of the Great Bear, is close to the zenith, while the main body of the constellation has swung round into the western sky.

Meteors

The **η-Aquariids** are one of the two meteor showers associated with Comet 1P/Halley (the other being the **Orionids**, in October). The **η-Aquariids** are not particularly favorably placed for northern-

hemisphere observers, because the radiant is near the celestial equator, near the 'Water Jar' in **Aquarius**, well below the horizon until late in the night (around dawn). However, meteors may still be seen in the eastern sky even when the radiant is below the horizon. There is a radiant map for the η-Aquariids on page 30.

Their maximum in 2021, on May 6, occurs when the Moon is a waning crescent, a few days after Last Quarter, so conditions are reasonably favorable. Maximum hourly rate is about 40 per hour and a large proportion (about 25 per cent) of the meteors leave persistent trains.

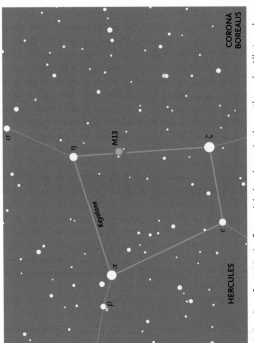

Finder chart for M13, the finest globular cluster in the northern sky. All stars down to magnitude 7.5 are shown.

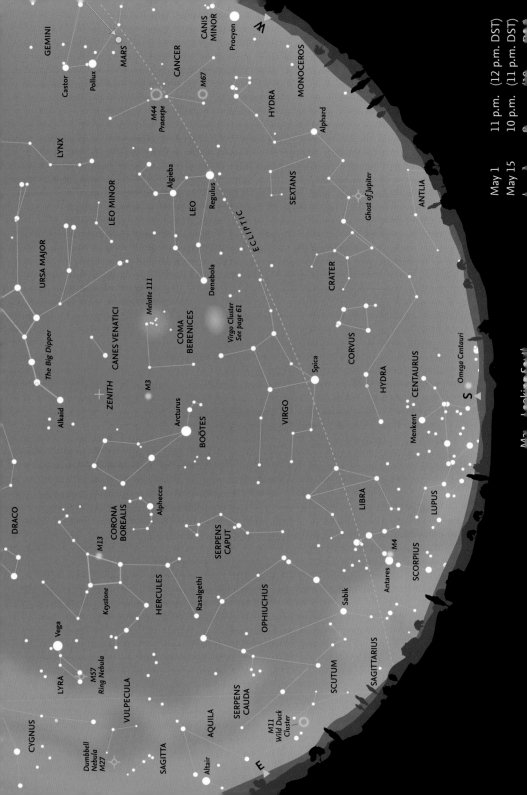

GEMINI
MARS
Castor
Pollux
CANIS MINOR
Procyon
CANCER
M67
M44
Praesepe
LYNX
HYDRA
MONOCEROS
Algieba
LEO MINOR
LEO
Regulus
Alphard
SEXTANS
URSA MAJOR
ECLIPTIC
Denebola
ANTLIA
Melotte 111
COMA BERENICES
Ghost of Jupiter
The Big Dipper
CANES VENATICI
Virgo Cluster
See page 61
CRATER
DRACO
Alkaid
ZENITH
M3
CORVUS
Arcturus
Spica
HYDRA
CORONA BOREALIS
BOOTES
VIRGO
CENTAURUS
Alphecca
M13
Menkent
Omega Centauri
SERPENS CAUT
Keystone
Rasalgethi
LIBRA
S
HERCULES
LUPUS
Vega
LYRA
OPHIUCHUS
Sabik
SCORPIUS
M57
Ring Nebula
Antares
M4
VULPECULA
SERPENS CAUDA
SCUTUM
SAGITTARIUS
CYGNUS
AQUILA
M11
Wild Duck
Cluster
Dumbell
Nebula
M27
SAGITTA
Altair
E

May 1   11 p.m. (12 p.m. DST)
May 15   10 p.m. (11 p.m. DST)

May – Looking South

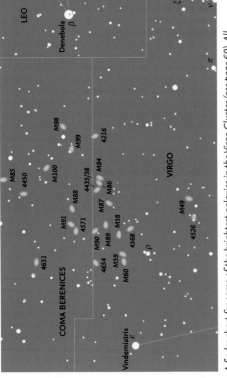

Early in the night, the constellation of *Virgo*, with *Spica* (α Virginis), lies due south, with *Leo* and both *Regulus* and *Denebola* (α and β Leonis, respectively) to its west still well clear of the horizon. The rather faint zodiacal constellation of *Libra* is now fully visible, together with most of *Scorpius* and ruddy *Antares* (α Scorpii). In the south, more of *Centaurus* may be seen, together with part of *Lupus*.

Virgo contains the nearest large cluster of galaxies, which is the center of the Local Supercluster, of which the Milky Way galaxy forms part. The Virgo Cluster contains some 2,000 galaxies, the brightest of which are visible in amateur telescopes.

Arcturus in *Boötes* is high in the south, with the distinctive circlet of *Corona Borealis* clearly visible to its east. The brightest star (α Coronae Borealis) is known as *Alphecca*. The large constellation of *Ophiuchus* (which actually crosses the ecliptic, and is thus the "13th" zodiacal constellation) is climbing into the eastern sky. Before the constellation boundaries were formally adopted by the International Astronomical Union in 1930, the southern region of Ophiuchus was regarded as

A finder chart for some of the brightest galaxies in the Virgo Cluster (see page 60). All stars brighter than magnitude 8.5 are shown.

forming part of the constellation of Scorpius, which had been part of the zodiac since antiquity.

The Moon's phases for May

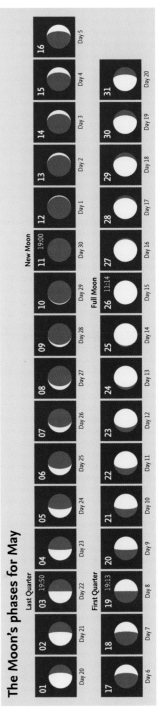

May – Moon and Planets

The Moon

The Moon is at its most distant apogee of the year (406,512 km) on May 11. On May 3, just before Last Quarter, the Moon is 4.2° south of **Saturn**, in **Capricornus**, and one day later it passes 4.6° south of **Jupiter** in **Aquarius**. On May 12–13, the Moon (immediately after New Moon, as a narrow waxing crescent) passes **Venus**, **Aldebaran** and then **Mercury**. A few days later (May 16–17) the Moon is visible 1.5° north of **Mars** in **Gemini** and then 3.1° south of **Pollux**. On May 19, the Moon passes between **Regulus** and **Algieba** (γ Leonis) in **Leo**. On May 23, waxing gibbous, the Moon is 6.5° north of **Spica**, and on May 26 (at Full Moon) there is a total lunar eclipse, visible from the Americas, the Pacific, Australia and parts of Asia. By May 30, the Moon is again south of **Saturn** in **Capricornus**.

The planets

Mercury is at greatest elongation east (22°) in the evening sky at mag. 0.3 on May 17. **Venus** (which passed superior conjunction in March), becomes visible for a brief period low in the evening sky at mag. -3.9. **Mars** (mag. 1.6–1.7) crosses **Gemini** over the month. **Jupiter** (mag. -2.2 to -2.4) is in **Aquarius**. **Saturn** (mag. 0.7–0.6) in **Capricornus** begins retrograde motion on May 24. **Uranus** (mag. 5.9) is in **Aries** and **Neptune** in **Aquarius** at mag. 7.9.

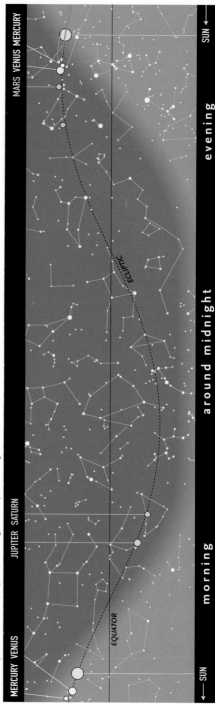

The path of the Sun and the planets along the ecliptic in May.

Calendar for May

03	16:58	Saturn 4.2°N of Moon
03	19:50	Last Quarter
04	21:02	Jupiter 4.6°N of Moon
06		η-Aquariid shower maximum
06	17:51	Neptune 4.4°N of Moon
10	21:06	Uranus 2.4°N of Moon
11	19:00	New Moon
11	21:53	Moon at apogee (406,512 km, greatest of year)
12	22:03	Venus 0.7°N of Moon
13	10:48	Aldebaran 5.5°S of Moon
13	17:59	Mercury 2.1°N of Moon
16	04:47	Mars 1.5°S of Moon
17	01:12	Pollux 3.1°N of Moon
17	05:54	Mercury at greatest elongation (22.0°E, mag. 0.3)
17	23:00 *	Venus 5.9°N of Aldebaran
19	17:59	Regulus 5.0°S of Moon
19	19:13	First Quarter
23	13:38	Spica 6.5°S of Moon
26	01:50	Moon at perigee (357,311 km)
26	11:14	Full Moon
26	11:18	Total lunar eclipse (Americas, Pacific, Australia, Asia)
26	17:25	Antares 4.8°S of Moon
29	06:00 *	Mercury 0.4°S of Venus
31	01:18	Saturn 4.2°N of Moon

*These objects are close together for an extended period around this time.

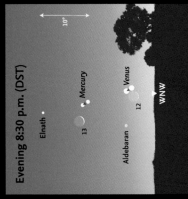

Early morning 5:30 a.m. (DST)

May 3–5 • The Moon is around Last Quarter when it passes Saturn and Jupiter in the southeast.

Evening 10 p.m. (DST)

May 15–16 • The Moon passes Mars, Castor and Pollux in the western sky. Mars has now faded to magnitude 1.6. Procyon is nearby, closer to the horizon.

Evening 8:30 p.m. (DST)

May 12–13 • After sunset, the narrow crescent Moon passes Venus, Aldebaran and Mercury, low in the west-northwest.

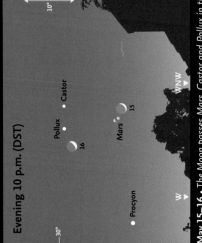

Evening 11:30 p.m. (BST)

May 18–19 • Still in the western sky, the First Quarter Moon passes between Regulus and Algieba.

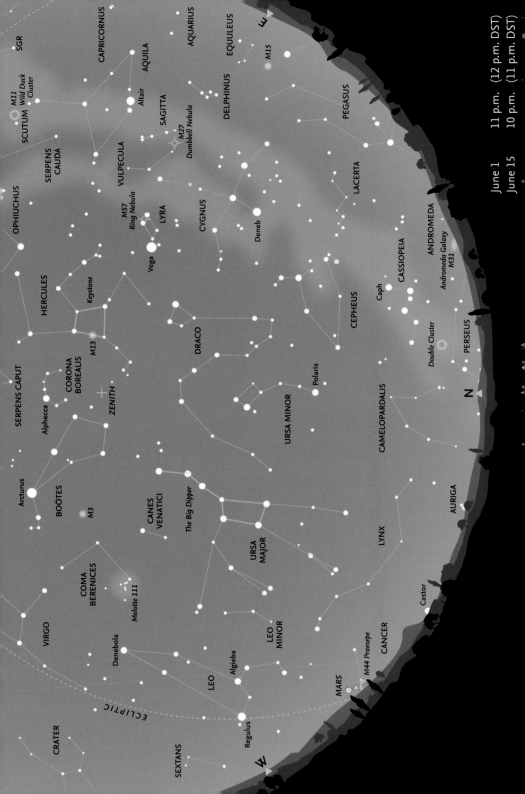

June – Looking North

With the approach of the summer solstice (June 21), twilight tends to persist in the northern United States and Canada, with four hours of darkness in Toronto in June, two hours in Seattle, and none in Vancouver. Even brighter stars, such as the seven stars making up the well-known asterism known as the **Big Dipper** in **Ursa Major** may be difficult to detect except around local midnight (1 a.m. DST). In the south of the United States, at New Orleans, Houston or Los Angeles, there are six or even seven hours of full darkness. Northern observers may have the compensation of seeing noctilucent clouds (NLC), electric-blue clouds visible during summer nights in the direction of the North Pole, for about a month or six weeks on either side of the solstice.

Two faint constellations, **Camelopardalis** and **Lynx**, may be glimpsed low on the northern and western horizons. **Leo** and **Regulus** (α Leonis) are heading downwards in the western sky, but **Cassiopeia**, farther east, is low, but clearly visible. The stars of the "Summer Triangle" (**Vega**, **Deneb** and **Altair**) in the constellations of **Lyra**, **Cygnus** and **Aquila**, respectively, are now readily seen in the east, and the small, distinctive constellation of **Delphinus** is well clear of the horizon.

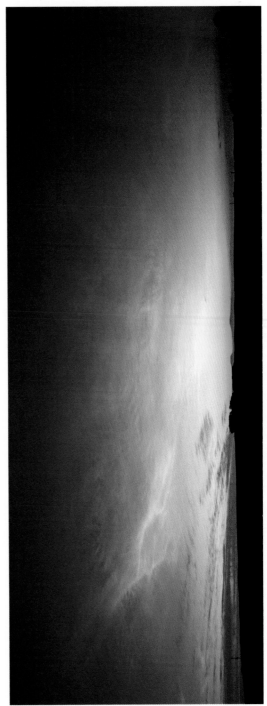

Noctilucent clouds, photographed on the night of 5–6 July 2016, from Tarbatness, Ross-shire, Scotland, by Denis Buczynski.

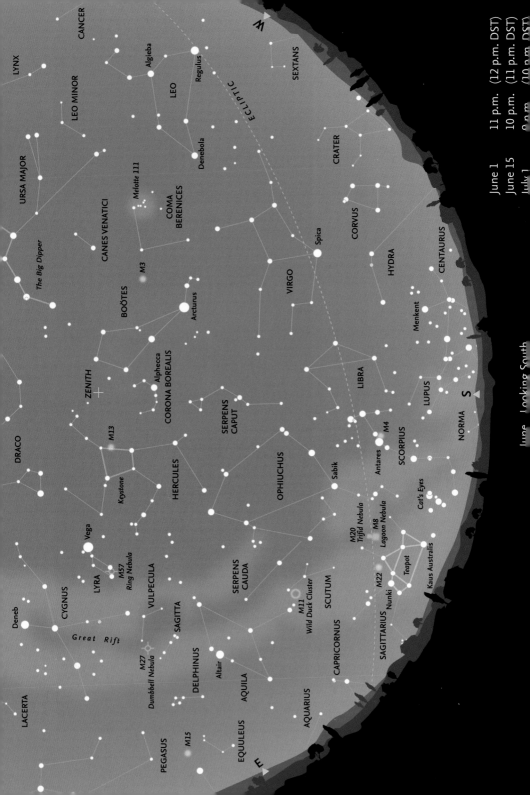

June Looking South

June 1	11 p.m.	(12 p.m. DST)
June 15	10 p.m.	(11 p.m. DST)
July 1	9 p.m.	(10 p.m. DST)

June – Looking South

The rather undistinguished constellation of *Libra* now lies almost due south. The red supergiant star *Antares* – the name means the "Rival of Mars" – in *Scorpius* is visible slightly to the east of the meridian, and even the "tail" or "sting" is visible with clear skies. Larger areas of *Centaurus* and *Lupus* are visible in the south, with *Sagittarius* to the east. Higher in the sky is the large constellation of *Ophiuchus* (the "Serpent Bearer"), lying between the two halves of *Serpens: Serpens Caput* ("Head of the Serpent") to the west and *Serpens Cauda* ("Tail of the Serpent") to the east. (Serpens is the only constellation to be divided into two distinct parts.) The ecliptic runs across Ophiuchus, and the Sun spends far more time in the constellation than it does in the "classical" zodiacal constellation of Scorpius, a small area of which lies between Libra and Ophiuchus.

Higher in the southern sky, the three constellations of *Boötes*, *Corona Borealis* and *Hercules* are now better placed for observation than at any other time of the year. This is an ideal time to observe the fine globular cluster of M13 in Hercules.

The Moon's phases for June

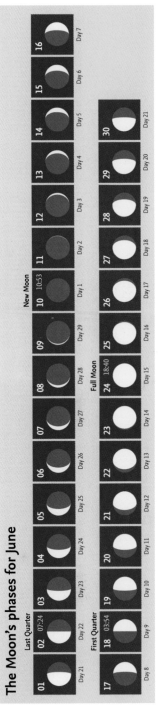

The constellations of Boötes and Corona Borealis are high in the sky in June. Arcturus has an orange tint and is the brightest star in the northern celestial hemisphere.

June – Moon and Planets

The Moon

One June 1, the day before Last Quarter, the Moon is 4.6° south of *Jupiter* in *Aquarius*. On June 10, at New Moon, there is an annular solar eclipse, visible from the high latitudes of Arctic Canada and Greenland (see page 20). On June 12–13, the Moon passes *Venus* and *Pollux* in *Gemini* and then 2.8° north of *Mars*. On June 15, the Moon is 5.0° north of *Regulus* in *Leo* and on June 19 it is close to *Spica* in *Virgo* as it sets in the west. On June 27 it passes, first 4.0° south of *Saturn* in *Capricornus* and then (on June 28), 4.5° south of *Jupiter* in *Aquarius*.

The planets

Mercury is invisible in twilight close to the Sun as it passes inferior conjunction on June 11. *Venus* is mag. -3.8, very close to the Sun, but may perhaps be glimpsed in the twilight at the end of the month. On June 22, it is south of *Pollux* and in the same field as Mars. At Last Quarter Moon, on June 2, *Mars* is 5.4° south of Pollux in *Gemini*. During the month, it moves from Gemini into *Cancer*, where it crosses the open cluster *Praesepe* on June 22 to 25 (see chart on page 23). *Jupiter*, in *Aquarius* (mag. -2.4 to -2.6) begins retrograde motion on June 21. *Saturn* (mag. 0.6–0.4) is moving westwards (retrograde) in Capricornus. *Uranus* (mag. 5.9–5.8) is in *Aries* and *Neptune* (mag. 7.9), in *Aquarius*, close to the border of *Pisces*, begins slow retrograde motion on June 27.

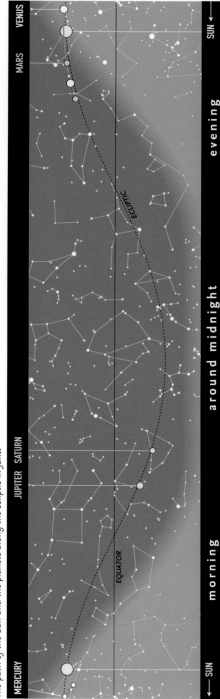

The path of the Sun and the planets along the ecliptic in June.

01	09:00	Jupiter 4.6°N of Moon
02	07:24	Last Quarter
02	14:00 *	Mars 5.4°S of Pollux
03	01:06	Neptune 4.5°N of Moon
07	06:16	Uranus 2.3°N of Moon
08	02:27	Moon at apogee (406,228 km)
09	16:51	Aldebaran 5.5°S of Moon
10	10:41	Annular solar eclipse (Arctic Canada, Greenland)
10	10:53	New Moon
10	13:09	Mercury 4.0°S of Moon
11	01:13	Mercury at inferior conjunction
12	06:42	Venus 1.5°S of Moon
13	06:52	Pollux 3.1°N of Moon
13	19:52	Mars 2.8°S of Moon
15	23:59	Regulus 5.0°S of Moon
18	03:54	First Quarter
19	22:09	Spica 6.5°S of Moon
21	03:32	Summer solstice
22	15:00 *	Venus 5.3°S of Pollux
23	03:56	Antares 4.8°S of Moon
23	09:55	Moon at perigee (359,956 km)
24	18:40	Full Moon
27	09:27	Saturn 4.0°N of Moon
28	18:42	Jupiter 4.5°N of Moon
30	09:09	Neptune 4.4°N of Moon

These objects are close together for an extended

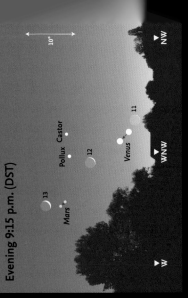

Evening 9:15 p.m. (DST)

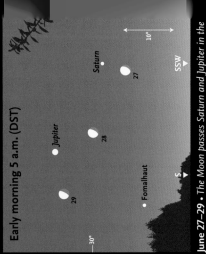

Evening 9 p.m. (DST)

June 2 • Mars with Castor and Pollux in the north-northwest. Mars has now faded to magnitude 1.7 (slightly fainter than Castor).

June 11–13 • Shortly after sunset the narrow crescent Moon passes Venus, Castor, Pollux and Mars, low in the west-northwest.

Early morning 5 a.m. (DST)

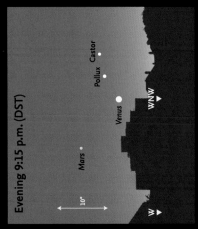

Evening 9:15 p.m. (DST)

June 22 • Venus with Castor and Pollux, close to the west-northwestern horizon. Mars (now at mag. 1.8)

June 27–29 • The Moon passes Saturn and Jupiter in the early morning, almost due south. Fomalhaut is closer to

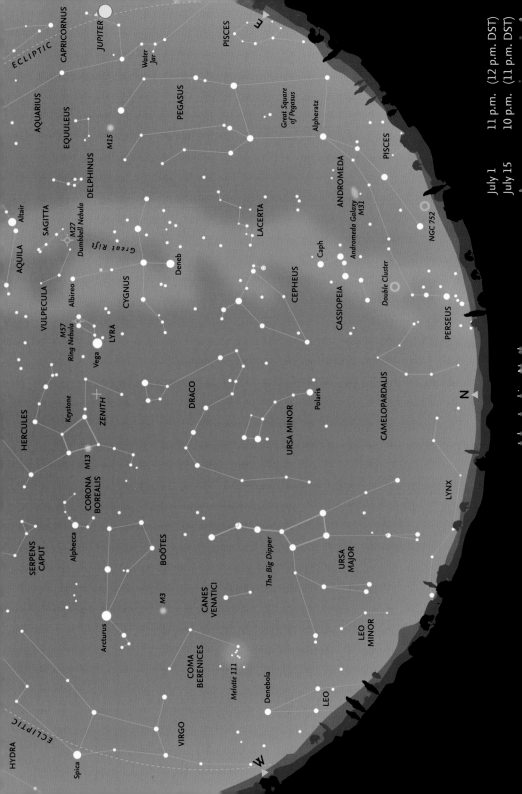

July – Looking North

As in June, light nights and the chance of observing noctilucent clouds persist throughout July, but later in the month (and particularly after midnight) some of the major constellations begin to be more easily seen. **Capella**, the brightest star in **Auriga** together with the rest of the constellation, is hidden below the northern horizon. **Cassiopeia** is clearly visible in the northeast and **Perseus**, to its south, is beginning to climb clear of the horizon. The band of the Milky Way, from Perseus through Cassiopeia toward **Cygnus**, stretches up into the northeastern sky. If the sky is dark and clear, you may be able to make out the small, faint constellation of **Lacerta**, lying across the Milky Way between Cassiopeia and Cygnus. In the east, the stars of **Pegasus** are now well clear of the horizon, with the main line of stars forming **Andromeda** roughly parallel to the horizon low in the northeast. **Alpheratz** (α Andromedae) is actually the star at the northeastern corner of the **Great Square of Pegasus. Cepheus** and **Ursa Major** are on opposite sides of **Polaris** and **Ursa Minor,** in the east and west, respectively. The head of **Draco** is very close to the zenith (which is in **Hercules**) so the whole of this winding constellation is readily seen.

Meteors

July brings increasing meteor activity, mainly because there are several minor radiants active in the constellations of **Capricornus** and **Aquarius.** Because of their location, however, observing conditions are not particularly favorable for northern-hemisphere observers, although the first shower, the **α-Capricornids,** active from July 2 to August 14 (peaking July 30, with a tail to August 14), does often produce very bright fireballs. The maximum rate, however, is only about 5 per hour. The parent body is Comet 169P/NEAT. The most prominent shower is probably that of the **Southern δ-Aquariids,** which are active from around July 13 to August 24, with a peak on July 30, although even then the rate is unlikely to reach 25 meteors per hour. In this case, the parent body may be Comet 96P/Machholz. This year, both shower maxima occur when the Moon is a waning gibbous, so observing conditions are not particularly favorable. A chart showing the **δ-Aquariids** radiant is shown on page 30. The **Perseids** begin on July 16 and peak on August 12–13.

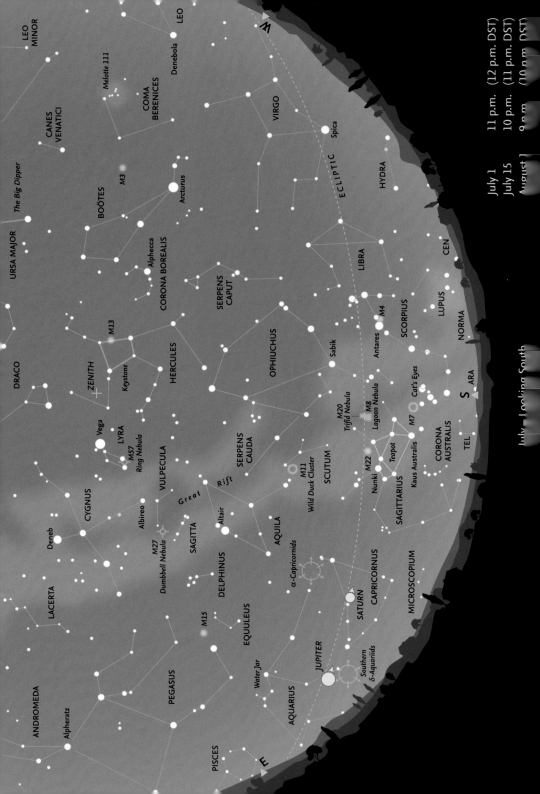

July, Looking South

July 1	11 p.m. (12 p.m. DST)
July 15	10 p.m. (11 p.m. DST)
August 1	9 p.m. (10 p.m. DST)

LEO MINOR
LEO
Denebola
CANES VENATICI
Melotte 111
COMA BERENICES
VIRGO
Spica
The Big Dipper
M3
BOÖTES
Arcturus
ECLIPTIC
HYDRA
URSA MAJOR
Alphecca
CORONA BOREALIS
SERPENS CAPUT
LIBRA
CEN
DRACO
M13
ZENITH
Keystone
HERCULES
OPHIUCHUS
Sabik
SCORPIUS
Antares
M4
LUPUS
NORMA
Vega
LYRA
M57
Ring Nebula
Cat's Eyes
M20
Trifid Nebula
M8
Lagoon Nebula
M7
ARA
S
Deneb
CYGNUS
Albireo
VULPECULA
SERPENS CAUDA
Great Rift
M11
Wild Duck Cluster
SCUTUM
Teapot
Nunki
M22
Kaus Australis
CORONA AUSTRALIS
TEL
LACERTA
M27
Dumbbell Nebula
SAGITTA
Altair
AQUILA
SAGITTARIUS
DELPHINUS
α-Capricornids
MICROSCOPIUM
ANDROMEDA
M15
EQUULEUS
CAPRICORNUS
SATURN
Alpheratz
PEGASUS
Water Jar
JUPITER
Southern δ-Aquarids
AQUARIUS
E
PISCES

July – Looking South

This is the best time of year to see **Scorpius**, with deep red **Antares** (α Scorpii), glowing just above the southern horizon. At around midnight, part of **Sagittarius**, with the distinctive asterism of the "Teapot," and the dense star clouds of the center of the Milky Way, are visible in the south, together with the small constellation of **Corona Australis**. The **Great Rift** – actually dust clouds that hide the more distant stars – runs down the Milky Way from Cygnus toward Sagittarius. Towards its northern end is the small constellation of **Sagitta** and the planetary nebula **M27** (the Dumbbell Nebula) in the otherwise insignificant constellation of **Vulpecula**. The sprawling constellation of **Ophiuchus** lies close to the meridian for a large part of the month, separating the two halves of the constellation of **Serpens**. The western half is called **Serpens Caput** (Head of the Serpent) and the eastern part **Serpens Cauda** (Tail of the Serpent). In the east, the bright **Summer Triangle**, consisting of **Vega** in **Lyra**, **Deneb** in **Cygnus** and **Altair** in **Aquila**, begins to dominate the southern sky, as it will throughout August and into September. The small constellation of Lyra, with Vega and a distinctive quadrilateral of stars to its east and south, lies not far from the zenith.

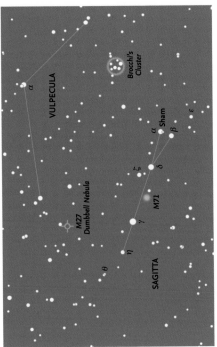

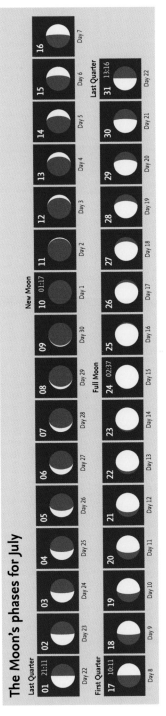

A finder chart for M27, the Dumbbell Nebula, a relatively bright (magnitude 8) planetary nebula – a shell of material ejected in the late stages of a star's lifetime – in the constellation of Vulpecula. All stars brighter than magnitude 7.5 are shown.

The Moon's phases for July

Last Quarter
01 21:11 — Day 22
02 — Day 23
03 — Day 24
04 — Day 25
05 — Day 26
06 — Day 27
07 — Day 28
08 — Day 29
09 — Day 30

New Moon
10 01:17 — Day 1
11 — Day 2
12 — Day 3
13 — Day 4
14 — Day 5
15 — Day 6
16 — Day 7

First Quarter
17 10:11 — Day 8
18 — Day 9
19 — Day 10
20 — Day 11
21 — Day 12
22 — Day 13
23 — Day 14

Full Moon
24 02:37 — Day 15
25 — Day 16
26 — Day 17
27 — Day 18
28 — Day 19
29 — Day 20
30 — Day 21

Last Quarter
31 13:16 — Day 22

July – Moon and Planets

The Earth

The Earth reaches aphelion, farthest from the Sun in its yearly orbit, at 22:27 on July 5, at a distance of 1.016729224 AU (152,100,527 km).

The Moon

The Moon is 5.6° north of **Aldebaran** in **Taurus** on July 6 and then passes between **Elnath** (β Tauri) and **Mercury**, which is low in the morning sky, on July 8. On July 12, the Moon passes north of **Venus** and **Mars** and then, on July 13, between **Regulus** and **Algieba** (γ Leonis). On July 17, at First Quarter, the Moon is 6.4° north of **Spica** in **Virgo**. On July 20, it passes 4.7° north of **Antares**. At Full Moon on July 24, the Moon is just 3.8° south of **Saturn**, very low in the south. The next day, July 25, the Moon is southwest of **Jupiter** and passes 4.2° south of the planet on July 26.

The planets

Mercury is at greatest western elongation (21.6°) low in the morning sky on July 4. It rapidly moves towards the Sun to superior conjunction in August. **Venus** is bright (mag. -3.9) but generally low in the evening sky. It leaves **Cancer**, entering **Leo**, and passes Mars in the middle of the month. **Mars**, in Leo, is not very bright (mag. 1.8) and is also low in the evening sky. **Jupiter** (mag. -2.6 to -2.8) is slowly retrograding in **Aquarius**. **Saturn** (mag. 0.4 to 0.2) is also slowly retrograding, in **Capricornus**. **Uranus** (mag. 5.8) is moving slowly eastwards in **Aries** and **Neptune** (mag. 7.8 to 7.9) is retrograding very slowly in **Aquarius**, near the border of **Pisces**. The minor planet **(6) Hebe** is at opposition (mag. 8.4) in **Aquila** on July 17 (see chart on page 26).

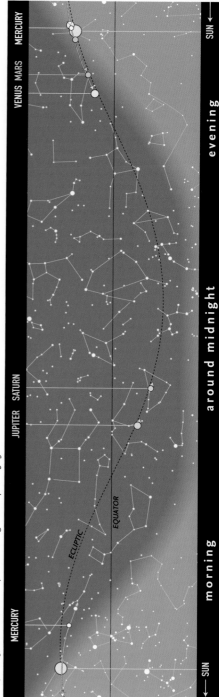

The path of the Sun and the planets along the ecliptic in July.

Calendar for July

01	21:11	Last Quarter
02–Aug.14		α-Capricornid meteor shower
04	15:25	Uranus 2.1°N of Moon
04	19:45	Mercury at greatest elongation (21.6°W, mag. 0.4)
05	14:47	Moon at apogee (405,341 km)
05	22:27	Earth at aphelion (1.016729224 AU = 152,100,527 km)
06	23:27	Aldebaran 5.6°S of Moon
08	04:38	Mercury 3.8°S of Moon
10	01:17	New Moon
10	12:58	Pollux 3.2°N of Moon
12	09:08	Venus 3.3°S of Moon
12	10:10	Mars 3.8°S of Moon
13–Aug.24		Southern δ-Aquariid meteor shower
13	05:33	Regulus 4.9°S of Moon
13	07:00 *	Mars 0.5°S of Venus
16–Aug.23		Perseid meteor shower
17	04:33	Spica 6.4°S of Moon
17	10:11	First Quarter
17	11:29	Minor Planet (6) Hebe at opposition (mag. 8.4)
20	12:38	Antares 4.7°S of Moon
21	10:24	Moon at perigee (364,520 km)
21	19:00 *	Venus 1.2°N of Regulus
24	02:37	Full Moon
24	16:39	Saturn 3.8°N of Moon
26	01:21	Jupiter 4.2°N of Moon
27	17:44	Neptune 4.2°N of Moon
29	16:00 *	Mars 0.7°N of Regulus
30		α-Capricornid shower maximum
30		Southern δ-Aquariid shower maximum
31	13:16	Last Quarter

These objects are close together for an extended period around this time.

Early morning 5 a.m. (DST)

July 7–8 • *In the early morning, shortly before sunrise, the narrow crescent Moon passes Aldebaran, Elnath and Mercury.*

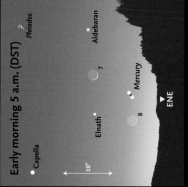

Evening 11 p.m. (DST)

July 23–25 • *The Moon is almost full when it passes Saturn low in the southeast. It is waning gibbous when it reaches Jupiter on July 25.*

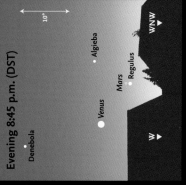

Evening 9 p.m. (DST)

July 11–13 • *The Moon passes Venus and Mars and then enters the constellation of Leo, the Lion. Mars is still at magnitude 1.8, but will not be hard to find, because it is close to Venus.*

Evening 8:45 p.m. (DST)

July 29 • *After sunset, Mars and Regulus are close together and form a nice triangle with Algieba and the much brighter Venus.*

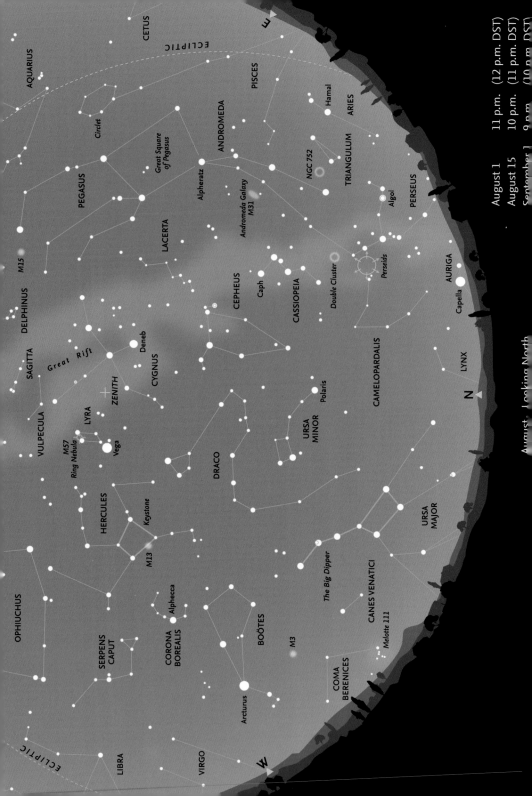

August. Looking North

August 1   11 p.m. (12 p.m. DST)
August 15   10 p.m. (11 p.m. DST)
September 1   9 p.m.   (10 p.m. DST)

August – Looking North

A brilliant Perseid fireball, streaking alongside the Great Rift in the Milky Way, photographed in 2012 by Jens Hackmann from near Weikersheim in Germany. Four additional, fainter Perseids are also visible in the image.

Ursa Major is now the "right way up" in the northwest, although some of the fainter stars in the south of the constellation are difficult to see. Beyond it, *Boötes* stands almost vertically in the west, but pale orange *Arcturus* is sinking toward the horizon. Higher in the sky, both *Corona Borealis* and *Hercules* are clearly visible.

In the northeast, *Capella* becomes visible later in the night, but most of *Auriga* still remains below the horizon. Higher in the sky, *Perseus* is gradually coming into full view and, later in the night and later in the month, the beautiful *Pleiades* cluster rises above the northeastern horizon. Between Perseus and *Polaris* lies the faint and unremarkable constellation of *Camelopardalis*.

Higher still, both *Cassiopeia* and *Cepheus* are well placed for observation, despite the fact that Cassiopeia is completely immersed in the band of the Milky Way, as is the "base" of Cepheus. *Pegasus* and *Andromeda* are now well above the eastern horizon and, below them, the constellation of *Pisces* is climbing into view. Two of the stars in the *Summer Triangle*, *Deneb* and *Vega*, are close to the zenith high overhead.

Meteors

August is the month when one of the best meteor showers of the year occurs: the *Perseids*. This is a long shower, generally beginning about July 16 and continuing until around August 23, with a maximum in 2021 on August 12–13, when the rate may reach as high as 100 meteors per hour (and on rare occasions, even higher). In 2021, maximum is just after New Moon, so conditions are favorable. The Perseids are debris from Comet 109P/Swift-Tuttle (the Great Comet of 1862). Perseid meteors are fast and many of the brighter ones leave persistent trains. Some bright fireballs also occur during the shower.

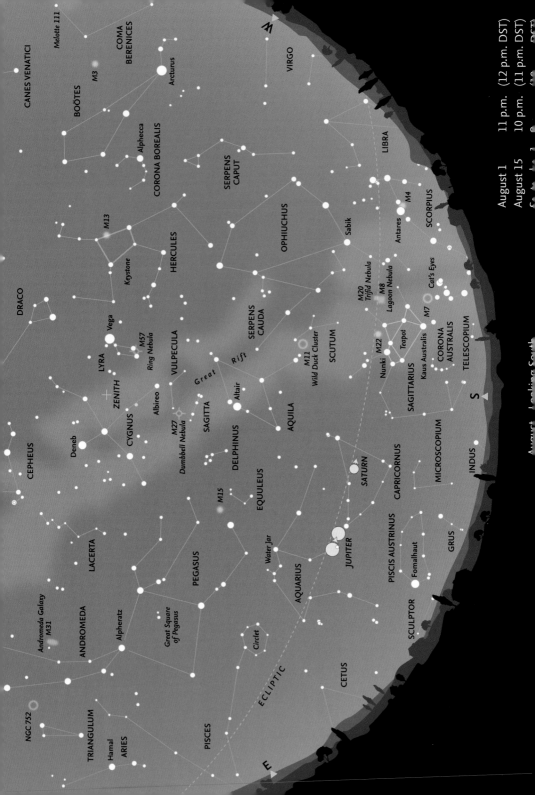

August – Looking South

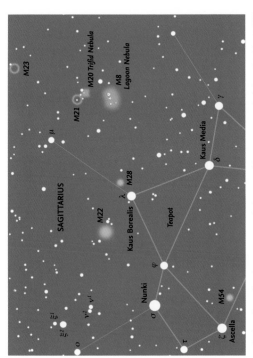

The whole stretch of the summer Milky Way stretches across the sky in the south, from **Cygnus,** high in the sky near the zenith, past **Aquila,** with bright **Altair** (α Aquilae), to part of the constellation of **Sagittarius** close to the horizon, where the pattern of stars known as the "Teapot" is visible. This area contains many nebulae and both open and globular clusters. Between **Albireo** (β Cygni) and Altair lie the two small constellations of **Vulpecula** and **Sagitta,** with the latter easier to distinguish (because of its shape) from the clouds of the Milky Way. Between Sagitta and **Pegasus** to the east lie the highly distinctive five stars that form the tiny constellation of **Delphinus** (again, one of the few constellations that actually bear some resemblance to the creatures after which they are named). Below Aquila, mainly in the star clouds of the Milky Way, lies **Scutum,** most famous for the bright open cluster, **M11** or the "Wild Duck Cluster," readily visible in binoculars. To the southeast of Aquila lie the two zodiacal constellations of **Capricornus** and **Aquarius** and, farther south, the constellation of **Piscis Austrinus** with the brilliant star **Fomalhaut.**

The Moon's phases for August

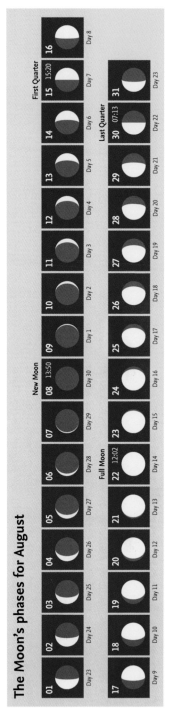

A finder chart for the gaseous nebulae M8 (the Lagoon Nebula), M20 (the Trifid Nebula) and the globular cluster M22, all in Sagittarius. Clusters M21, M23 (open) and M28 (globular) are faint. The chart shows all stars brighter than magnitude 7.5.

August – Moon and Planets

The Moon

The Moon passes north of **Aldebaran** on August 3. On August 6, the waning crescent is between **Castor** (α Geminorum) and **Alhena** (γ Geminorum), later that day it is 3.1° south of **Pollux**. On August 10, just after New Moon (on August 8), the hair-thin waxing crescent passes **Mars**, low in the evening sky, and the next day 4.3° north of **Venus**. On August 13, the Moon passes **Spica** in **Virgo**, and is 4.5° north of **Antares** in **Scorpius** on August 16. On August 20, the Moon is 3.7° south of **Saturn** and the next day 4.0° south of **Jupiter**. It passes **Aldebaran** in **Taurus** on August 30.

The planets

Mercury is at superior conjunction on the far side of the Sun on August 1. **Venus** is in **Virgo**, but too low to be readily visible except briefly in the evening twilight. **Mars** is in **Leo**, initially close to **Regulus**, but moving eastward toward **Virgo**, and very low in the sky. **Jupiter** (mag. -2.8 to -2.9) is in **Capricornus** and comes to opposition on August 20. **Saturn** (mag. 0.2) is at opposition in **Capricornus** on August 2. **Uranus** (mag. 5.8–5.7) in **Aries** begins retrograde motion on August 20. **Neptune** (mag. 7.9) remains in **Aquarius**.

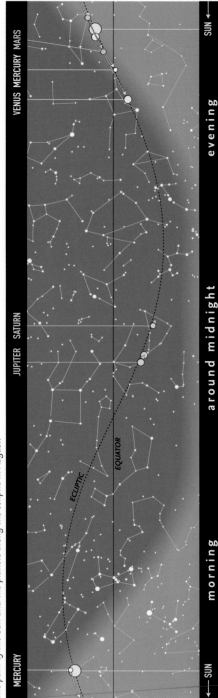

The path of the Sun and the planets along the ecliptic in August.

Calendar for August

01	00:29	Uranus 1.8°N of Moon
01	14:07	Mercury at superior conjunction
02	06:14	Saturn at opposition (mag. 0.2)
02	07:35	Moon at apogee (404,410 km)
03	06:53	Aldebaran 5.7°S of Moon
06	20:14	Pollux 3.1°N of Moon
08	13:50	New Moon
09	03:19	Mercury 3.4°S of Moon
09	12:04	Regulus 4.8°S of Moon
10	00:41	Mars 4.3°S of Moon
11	06:59	Venus 4.3°S of Moon
12–13		Perseid shower maximum
13	10:03	Spica 6.1°S of Moon
15	15:20	First Quarter
16	19:06	Antares 4.5°S of Moon
17	09:16	Moon at perigee (369,124 km)
19	04:00 *	Mercury 0.1°S of Mars
20	00:28	Jupiter at opposition (mag. −2.9)
20	22:15	Saturn 3.7°N of Moon
22	04:56	Jupiter 4.0°N of Moon
22	12:02	Full Moon
24	01:55	Neptune 4.0°N of Moon
28–Sep.05		α-Aurigid meteor shower
28	08:57	Uranus 1.5°N of Moon
30	02:22	Moon at apogee (404,100 km)
30	07:13	Last Quarter
30	14:55	Aldebaran 6.0°S of Moon

* These objects are close together for an extended period around this time.

Early morning 5:30 a.m. (DST)

August 6–7 • *Shortly before sunrise, the narrow crescent Moon passes Castor and Pollux. Alhena is farther east.*

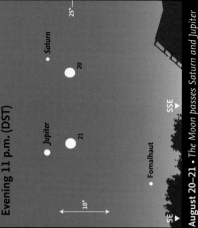

Evening 8:30 p.m. (DST)

August 9–11 • *The waxing, but still very narrow, crescent Moon passes Mars and Venus low in the west. Mars is not very bright (mag. 1.8) and may not be easy to detect*

Evening 9 p.m. (DST)

August 15–16 • *The Moon passes north of Antares. Sabik is about ten degrees higher and the Cat's Eyes are due south, closer to the horizon.*

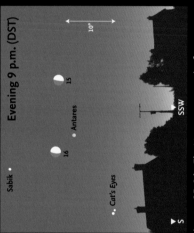

Evening 11 p.m. (DST)

August 20–21 • *The Moon passes Saturn and Jupiter in the south-southeast. Fomalhaut is closer to the horizon and a little more to the east.*

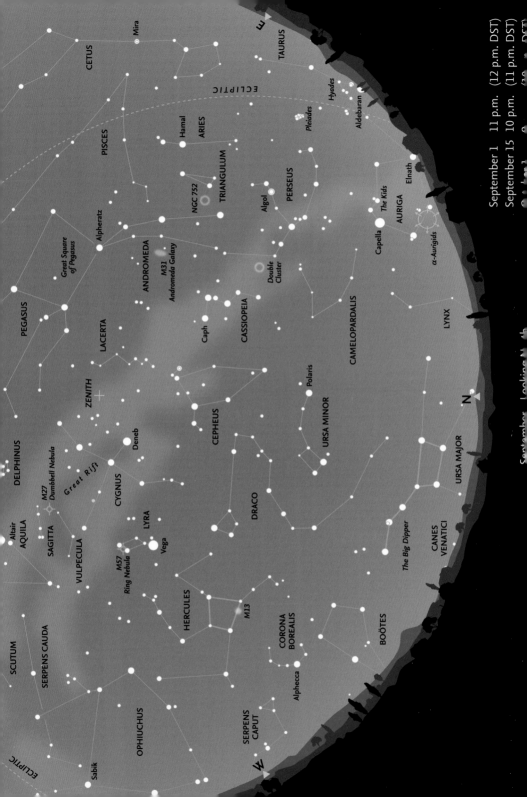

September 1   11 p.m.   (12 p.m. DST)
September 15   10 p.m.   (11 p.m. DST)

September – Looking North

The (northern) autumnal equinox occurs on September 22, when the Sun moves south of the equator in Virgo.

Ursa Major is now low in the north, and to the northwest **Arcturus** and much of **Boötes** have sunk below the horizon. In the northeast, **Auriga** is beginning to climb higher in the sky. Later in the month, **Taurus**, with orange **Aldebaran** (α Tauri), and even **Gemini**, with **Castor** and **Pollux**, become visible in the east and northeast. Due east, **Andromeda** is now easily visible, with the small constellations of **Triangulum** and **Aries** (the latter a zodiacal constellation) directly below it. Practically the whole of the northern Milky Way is visible, arching across the sky, both in the north and in the south. It is not particularly clear in Auriga, or even **Perseus**, but in **Cassiopeia** and on toward **Cygnus** the clouds of stars become easier to see. The **Double Cluster** in Perseus is well placed for observation. **Cepheus** is "upside-down" near the zenith, and the head of **Draco** and **Hercules** beyond it are well placed for observation.

Meteors

After the major Perseid shower in August, there is very little shower activity in September. One minor, but very extended, shower, known as the **α-Aurigids**, tends to have two peaks of activity. The principal peak occurs on September 1. In 2021, the Moon is a waning crescent, so conditions are reasonably favorable. At maximum, however, the hourly rate hardly reaches 10 meteors per hour, although the meteors are bright and relatively easy to photograph. Activity from this shower may even extend into October. The **Southern Taurid** shower begins this month (on September 10) and, although rates are low, often produces very bright fireballs. This is a very long shower, lasting until about November 20. As a slight compensation for the lack of shower activity, however, in September the number of sporadic meteors reaches its highest rate at any time during the year.

The twin open clusters, known as the Double Cluster, in Perseus (more formally called h and χ Persei) are close to the main portion of the Milky Way.

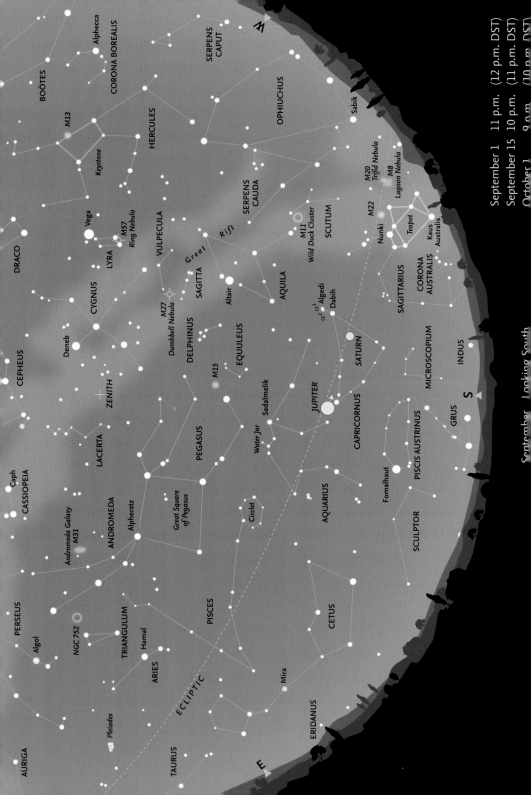

September 1   11 p.m. (12 p.m. DST)
September 15 10 p.m. (11 p.m. DST)
October 1   9 p.m. (10 p.m. DST)

September Looking South

September – Looking South

The **Summer Triangle** is now high in the southwest, with the Great Square of **Pegasus** high in the southeast. Below Pegasus are the two zodiacal constellations of **Capricornus** and **Aquarius.** In what is otherwise an unremarkable constellation, **Algedi** (α Capricorni) is actually a visual binary, with the two stars (α¹ Cap and α² Cap) readily seen with the naked eye. **Dabih** (β Capricorni), just to the south, is also a double star, and the components are relatively easy to separate with binoculars. In Aquarius, just to the east of **Sadalmelik** (α Aquarii) there is a small asterism consisting of four stars, resembling a tiny letter "Y," known as the "Water Jar." Below Aquarius is a sparsely populated area of the sky with just one bright star in the constellation of **Piscis Austrinus.** In classical illustrations, water is shown flowing from the "Water Jar" toward bright **Fomalhaut** (α Piscis Austrini).

Another zodiacal constellation, **Pisces,** is now clearly visible to the east of Aquarius. Although faint, there is a distinctive asterism of stars, known as the "Circlet," south of the Great Square and another line of faint stars to the east of Pegasus. Still farther down toward the horizon

The zodiacal constellation of Capricornus, photographed in 2009. The brilliant object is the planet Jupiter, which will again be in the constellation for part of 2021.

is the constellation of **Cetus,** with the famous variable star **Mira** (o Ceti) at its center. When Mira is at maximum brightness (around mag. 3.5) it is clearly visible to the naked eye, but it disappears as it fades toward minimum (about mag. 9.5 or less). There is a finder chart for Mira on page 97.

The Moon's phases for September

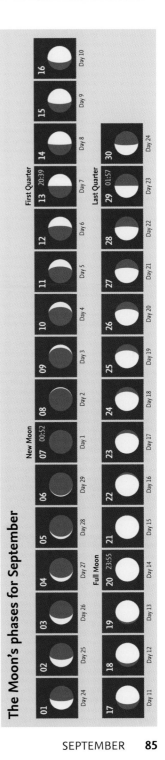

September – Moon and Planets

The Moon

On September 3, the Moon passes 3.0° south of *Pollux* in *Gemini*. On September 5, the Moon is 4.8° north of *Regulus* in the morning sky. On September 9, the young crescent Moon is 5.9° north of *Spica*, but the star will be difficult to detect in the evening twilight. The next day, the Moon passes 4.1° north of *Venus*. On September 17–18, the Moon passes south of first *Saturn*, and then *Jupiter* (both in *Capricornus*). On September 25, the Moon passes south of the *Pleiades* and then, on September 26, is 6.2° north of *Aldebaran* in *Taurus*. On September 30, the waning crescent Moon is 2.8° south of *Pollux* in *Gemini*.

The planets

Mercury is at greatest eastern elongation 26.8° on September 14, but only mag. 0.1. Initially, *Venus* is close to *Spica* in *Virgo* but then moves into *Libra*. Generally, it is in the evening twilight, but becomes slightly higher towards the end of the month. *Mars* is too close to the Sun to be seen. *Jupiter* (mag. -2.9 to -2.7) is still retrograding slowly in *Capricornus*, close to *Deneb Algedi* (δ Capricorni). *Saturn* (mag. 0.3–0.5) is also retrograding extremely slowly in *Capricornus*. *Uranus* (mag. 5.7) remains in *Aries* and *Neptune* (mag. 7.8) comes to opposition in *Aquarius* on September 14 (see the chart on page 25). Minor planet (2) *Pallas* (mag. 8.5) is at opposition in western *Pisces* on September 11 (see chart on page 27).

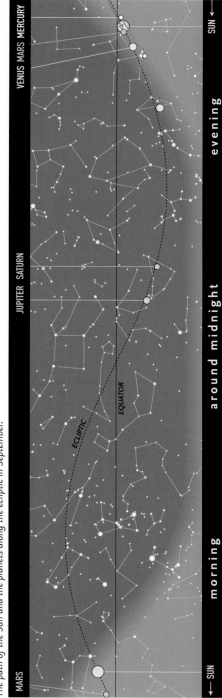

The path of the Sun and the planets along the ecliptic in September.

Calendar for September

01		α-Aurigid shower maximum
03	04:37	Pollux 3.0°N of Moon
05		Venus 1.7°N of Spica
05	06:00 *	Regulus 4.8°S of Moon
07	20:16	New Moon
07	00:52	Mars 4.2°S of Moon
07	16:21	Mercury 6.5°S of Moon
08	20:19	Spica 5.9°S of Moon
09	16:32	Southern Taurid meteor shower
10–Nov.20		Venus 4.1°S of Moon
10	02:09	Minor planet (2) Pallas at
11	01:48	opposition (mag. 8.5)
11	10:03	Moon at perigee (368,461 km)
13	00:31	Antares 4.2°S of Moon
13	20:39	First Quarter
14	04:24	Mercury at greatest elongation
		(26.8°E, mag. 0.1)
14	09:21	Neptune at opposition (mag. 7.8)
17	02:33	Saturn 3.8°N of Moon
18	06:54	Jupiter 4.0°N of Moon
20	08:45	Neptune 4.0°N of Moon
20	23:55	Full Moon
22	19:21	Autumnal equinox
24	16:08	Uranus 1.4°N of Moon
26	21:44	Moon at apogee (404,640 km)
26	22:54	Aldebaran 6.2°S of Moon
29	01:57	Last Quarter
30	13:18	Pollux 2.8°N of Moon

* These objects are close together for an extended period around this time.

Evening 7:45 p.m. (DST)

September 5 • *After sunset, Spica is less than two degrees below the much brighter Venus.*

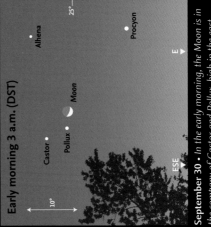

Evening 7:45 p.m. (DST)

September 8–9 • *The waxing crescent Moon passes Mercury (mag. 0.3), Spica (mag. 1.0) and Venus (mag. –3.9), after sunset near the western horizon.*

Evening 9 p.m. (DST)

September 16–17 • *The waxing gibbous Moon passes Saturn and Jupiter. Fomalhaut is close to the horizon in the southeast.*

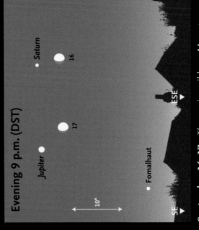

Early morning 3 a.m. (DST)

September 30 • *In the early morning, the Moon is in the company of Castor and Pollux, high in the east. Alhena and Procyon are nearby.*

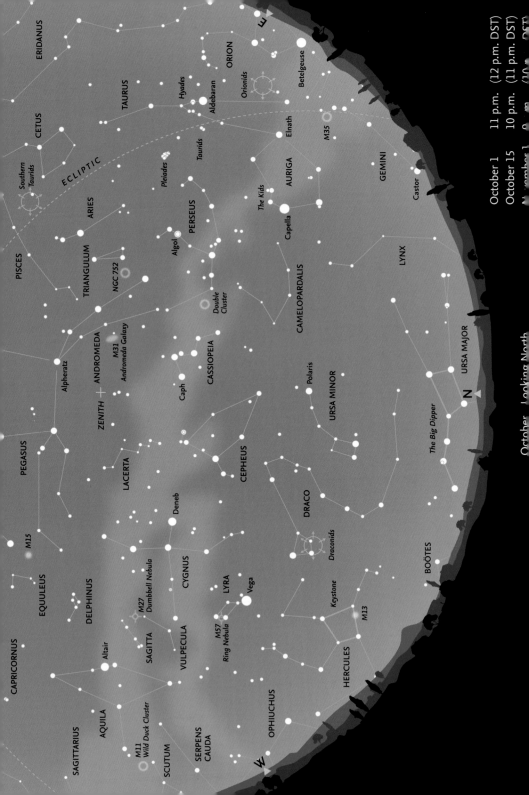

October 1   11 p.m. (12 p.m. DST)
October 15   10 p.m. (11 p.m. DST)
November 1   9 p.m. (10 p.m. DST)

October Looking North

October – Looking North

Ursa Major is grazing the horizon in the north, while high overhead are the constellations of *Cepheus, Cassiopeia* and *Perseus*, with the Milky Way between Cepheus and Cassiopeia near the zenith. *Auriga* is now clearly visible in the east, as is *Taurus* with the *Pleiades, Hyades* and orange *Aldebaran*. Also in the east, *Orion* and *Gemini* are starting to rise clear of the horizon.

The constellations of *Boötes* and *Corona Borealis* are now lost to view in the northwest, and *Hercules* is also descending toward the western horizon. The three stars of the Summer Triangle are still clearly visible, although *Aquila* and *Altair* are beginning to approach the horizon in the west. Toward the end of the month (October 31) Summer Time ends in Europe. In North America the end of DST comes next month, in November.

Meteors

The *Orionids* are the major, fairly reliable meteor shower active in October. Like the May *η-Aquariid* shower, the Orionids are associated with Comet 1P/Halley. During this second pass through the stream of particles from the comet, slightly fewer meteors are seen than in May, but conditions are more favorable for northern observers. In both showers the meteors are very fast, and many leave persistent trains. Although the Orionid maximum is quoted as October 21, in fact there is a very broad maximum, lasting about a week roughly centered on that date, with hourly rates around 25. Occasionally, rates are higher (50–70 per hour). In 2021, Full Moon is on October 20, so there will be considerable interference from moonlight.

The faint shower of the *Southern Taurids* (often with bright fireballs) peaks on October 10. The Southern Taurid maximum occurs when the Moon is a waxing crescent, so conditions are more favorable than

The constellation of Perseus is not only the location of the radiant of the Perseid meteor shower in August, but is also well known for the pair of open clusters, called the Double Cluster (near the top edge of the image), close to the border with Cassiopeia, and also for Algol (β Persei), the famous variable star (near the center of the photograph). The Pleiades, an open cluster in Taurus, is close to the bottom edge of the image.

for the Orionids. Towards the end of the month (around October 20), another shower (the Northern Taurids) begins to show activity, which peaks early in November. The parent comet for both Taurid showers is Comet 2P/Encke. The meteors in both Taurid streams are relatively slow and bright.

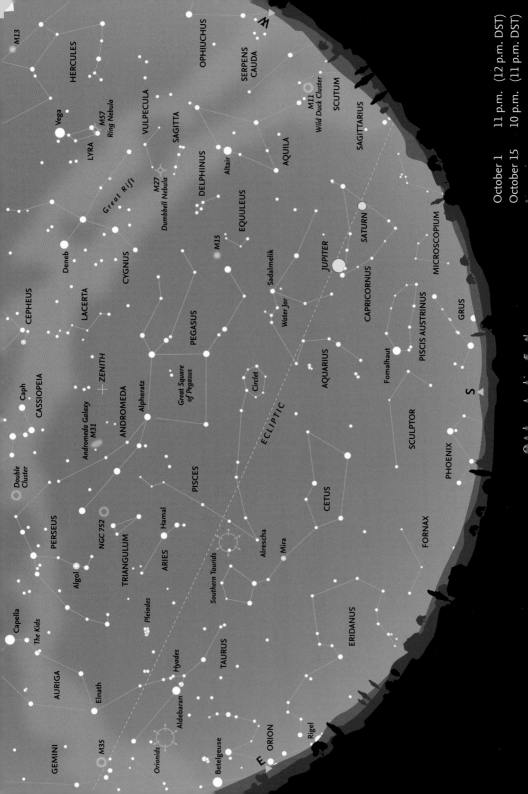

October 1   11 p.m.   (12 p.m. DST)
October 15   10 p.m.   (11 p.m. DST)

October – Looking South

The Great Square of **Pegasus** dominates the southern sky, framed by the two chains of stars that form the constellation of **Pisces**, together with **Alrescha** (α Piscium) at the point where the two lines of stars join. Also clearly visible is the constellation of **Cetus**, below Pegasus and Pisces. Although **Capricornus** is now lower, **Aquarius** to its east is well placed in the south, with solitary **Fomalhaut** and the constellation of **Piscis Austrinus** beneath it. The inconspicuous constellation of **Sculptor** appears in the south, as well as parts of **Grus**, **Phoenix** and, farther east, the northernmost stars of **Eridanus**.

The main band of the Milky Way and the Great Rift runs down from **Cygnus**, through **Vulpecula**, **Sagitta** and **Aquila** toward the western horizon. **Delphinus** and the tiny, unremarkable constellation of **Equuleus** lie between the band of the Milky Way and Pegasus. **Andromeda** is clearly visible high in the sky to the southeast, with the small constellation of **Triangulum** and the zodiacal constellation of **Aries** below it. **Perseus** is high in the east, and by now the **Pleiades** and **Taurus** are well clear of the horizon. Later in the night, and later in the month, **Orion** rises in the east, a sign that the autumn season has arrived and of the steady approach of winter.

The Moon's phases for October

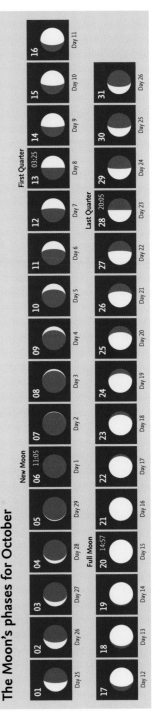

The constellation of Aquarius is one of the constellations that is visible in late summer and early autumn. The four stars forming the "Y"-shape of the "Water Jar" may be seen to the east of Sadalmelik (α Aquarii), the brightest star (top center).

October – Moon and Planets

The Moon

On October 3, the waning crescent Moon passes between **Regulus** (α Leonis) and **Algieba** (γ Leonis) in the early morning sky. On October 9, the Moon is 2.9° north of **Venus**, and it passes 4.0° north of **Antares** on October 10, both low in the evening sky. On October 14, the Moon passes 3.9° south of **Saturn** and, the next day, 4.1° south of **Jupiter**. The waning gibbous Moon is 6.4° north of **Aldebaran** on October 24. On October 27, the Moon passes 3.0° south of **Pollux**, and on October 30 passes 5.1° north of **Regulus** in **Leo**.

The planets

Mercury passes inferior conjunction on October 9 and is mag. -0.6 at greatest western elongation (18.4°) on October 25. **Venus** is close to **Antares** for several days low in the eastern sky, passing 1.5° due north of the star on October 16. It is very bright (mag. -4.5) when at its greatest possible eastern elongation (47°) on October 29. **Mars** is at superior conjunction on October 8. **Jupiter** (mag. -2.5 to -2.7) is initially retrograding in **Capricornus** but begins direct motion on October 30. **Saturn** (mag. 0.5–0.6), also in **Capricornus**, reverts to direct motion on October 19. **Uranus** (mag. 5.7) is retrograding in **Aries**. **Neptune** (mag. 7.8) continues retrograde motion in **Aquarius**.

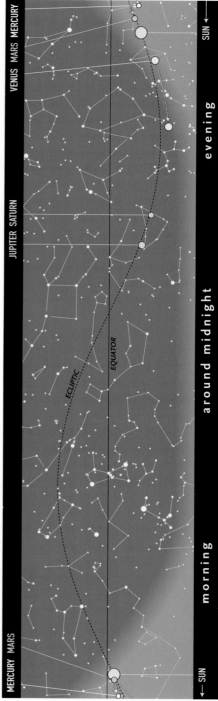

The path of the Sun and the planets along the ecliptic in October.

Calendar for October

*These objects are close together for an extended

Morning 6 a.m. (DST)

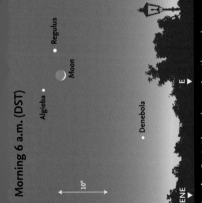

10°

Denebola •

Algieba •

Moon

• Regulus

ENE

E

October 3 • In the early morning, the waning crescent Moon is close to Regulus and Algieba. Denebola is closer to the horizon and more to the northeast.

Evening 7 p.m. (DST)

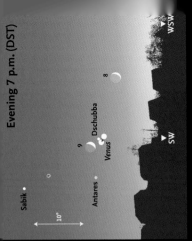

10°

Jupiter

Moon

Saturn

SSE

October 14 • The Moon with Saturn and Jupiter in the south-southeast

Evening 7 p.m. (DST)

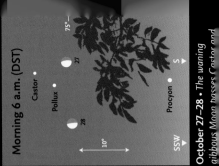

10°

Sabik •

Sabik •

Antares •

Dschubba

Venus

WSW

SW

8

9

October 8–9 • Shortly after sunset, the crescent Moon passes Venus which is close to Dschubba (δ Sco). On October 16, Venus will have reached Antares (see below).

Evening 7 p.m. (DST)

10°

Sabik •

Venus

Antares

SW

October 16 • Antares is 1.5° below the much brighter Venus

Morning 6 a.m. (DST)

Castor •
Pollux •

27
28

75°—

Procyon •

10°

S

SSW

October 27–28 • The waning gibbous Moon passes Castor and

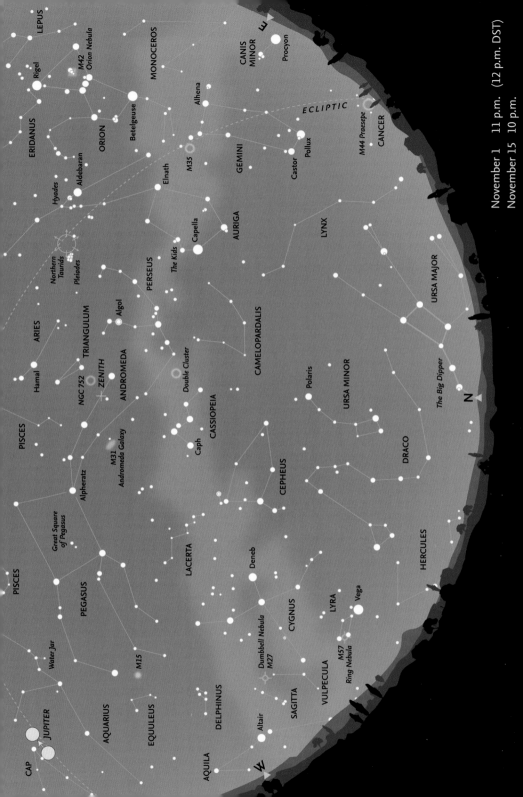

November – Looking North

The constellation of Auriga, with brilliant Capella (mag. 0.08), which, although appearing as a single star, is actually a quadruple system, consisting of a pair of yellow giant stars, gravitationally bound to a more distant pair of red dwarfs.

Most of **Aquila** has now disappeared below the horizon, but two of the stars of the Summer Triangle, **Vega** in **Lyra** and **Deneb** in **Cygnus**, are still clearly visible in the west. The head of **Draco** is now low in the northwest and only a small portion of **Hercules** remains above the horizon. The southernmost stars of **Ursa Major** are now coming into view. The Milky Way arches overhead, with the denser star clouds in the west and the less heavily populated region through **Auriga** and **Monoceros** in the east. High overhead, **Cassiopeia** is near the zenith and **Cepheus** has swung round to the northwest, while Auriga is now high in the northeast. **Gemini**, with **Castor** and **Pollux**, is well clear of the eastern horizon, and even **Procyon** (α Canis Minoris) is just climbing into view almost due east.

Meteors

The **Northern Taurid** shower, which began in mid-October, reaches maximum – although with just a low rate of about five meteors per hour – on November 11. First Quarter is on November 12, so conditions are reasonably favorable. The shower gradually trails off, ending around December 10. There is an apparent 7-year periodicity in fireball activity, but 2021 is unlikely to be a peak year. Far more striking, however, are the **Leonids**, which have a relatively short period of activity (November 5–29), with maximum on November 17–18. This shower is associated with Comet 55P/Tempel-Tuttle and has shown extraordinary activity on various occasions with many thousands of meteors per hour. High rates were seen in 1999, 2001 and 2002 (reaching about 3,000 meteors per hour) but have fallen dramatically since then. The rate in 2021 is likely to be about 15 per hour. These meteors are the fastest shower meteors recorded (about 70 km per second) and often leave persistent trains. The shower is very rich in faint meteors. In 2021, maximum is just before Full Moon (on November 19), so conditions are very unfavorable.

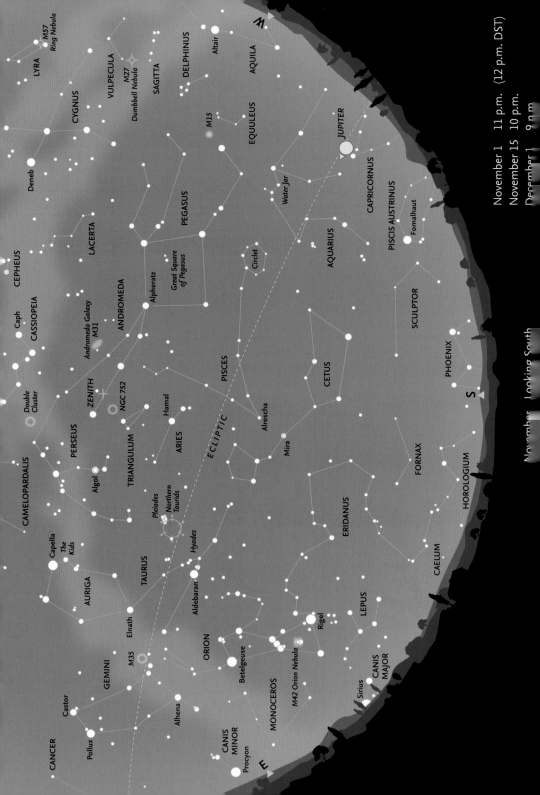

November 1   11 p.m. (12 p.m. DST)
November 15   10 p.m.
December 1   9 p.m.

November   Looking South

November – Looking South

Orion has now risen above the eastern horizon, and much of the long, straggling constellation of **Eridanus** (which begins near **Rigel**) is visible to the west of Orion as is the small constellation of **Fornax**. Parts of **Horologium** and **Phoenix** are peeping above the southern horizon. Higher in the sky, **Taurus**, with the **Pleiades** cluster, and orange **Aldebaran** are now easy to observe. To their west, both **Pisces** and **Cetus** are close to the meridian. The famous long-period variable star, **Mira** (o Ceti), with a typical range of magnitude 3.4 to 9.8, is favorably placed for observation. In the southwest, **Capricornus** has slipped below the horizon, but **Aquarius** remains visible. Even farther west, **Altair** may be seen early in the night, but most of **Aquila** has already disappeared from view. **Delphinus**, together with **Sagitta** and **Vulpecula** in the Milky Way, will soon vanish for another year. Both **Pegasus** and **Andromeda** are easy to see, and one of the lines of stars that make up Andromeda finishes close to the zenith, which is also close to one of the outlying stars of Perseus, high in the east.

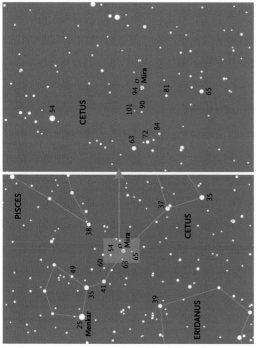

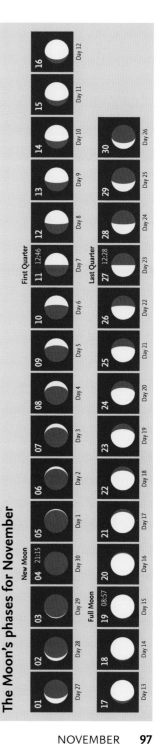

Finder and comparison charts for Mira (o Ceti). The chart on the left shows all stars brighter than magnitude 6.5. The chart on the right shows stars down to magnitude 10.0. The comparison star magnitudes are shown without the decimal point.

The Moon's phases for November

November – Moon and Planets

The Moon

On November 3, the extremely narrow waning crescent Moon (one day before New Moon) passes close to **Mercury** and **Spica**. On November 7–8, the young waxing crescent may be glimpsed low in the evening sky, close to **Venus**. On November 10, the Moon passes 4.1° south of **Saturn** and then on November 11, 4.4° south of **Jupiter**. On November 19, at Full Moon, there is a partial lunar eclipse. Just the start of this eclipse is visible from the UK, but it is fully visible from North and South America, the Pacific region, Australia and most of Asia. The next day (November 20), the Moon is north of **Aldebaran**. On November 24 it is 2.5° south of **Pollux**, and then on November 26, 5.2° north of **Regulus** in **Leo**. By November 30, it is north of **Spica** in **Virgo**.

The planets

Mercury is too close to the Sun to be visible during the month. It is at superior conjunction on November 29. **Venus** is low in **Sagittarius**. It passes south of **Nunki** (σ Sagattarii) on November 19. **Mars** (mag. 1.6) is in **Virgo**, moving into **Libra**, and may be glimpsed in the morning twilight. **Jupiter** (mag. -2.5 to -2.3) is in **Capricornus**, near **Deneb Algedi** (δ Capricorni). **Saturn** (mag. 0.6–0.7) is also in **Capricornus**. **Uranus** (mag. 5.7) is in **Aries** and comes to opposition on November 4. **Neptune** (mag. 7.8) continues slow retrograde motion in **Aquarius**. Dwarf planet **(1) Ceres** is at opposition at mag. 7.0 in **Taurus**, not that far from **Aldebaran**, on November 27 (see the chart on page 27).

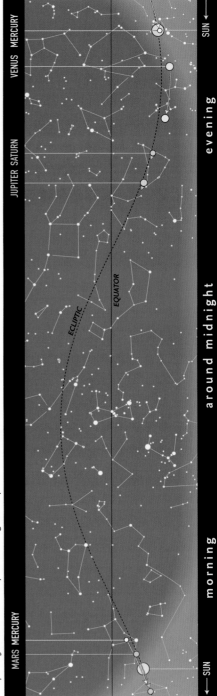

The path of the Sun and the planets along the ecliptic in November.

Date	Time	Event
01	02:00 *	Mercury 4.4°N of Spica
03	11:49	Spica 5.8°S of Moon
03	18:39	Mercury 1.2°S of Moon
04	04:36	Mars 2.3°S of Moon
04	21:15	New Moon
04	23:58	Uranus at opposition (mag. 5.7)
05	22:18	Moon at perigee (358,844 km)
05–29		Leonid meteor shower
06	15:58	Antares 3.9°S of Moon
07		North American Daylight Saving Time ends
08	05:21	Venus 1.1°S of Moon
10	04:00 *	Mercury 1.1°N of Mars
10	14:24	Saturn 4.1°N of Moon
11	12:46	First Quarter
11	17:16	Jupiter 4.4°N of Moon
12		Northern Taurid shower maximum
13	18:37	Neptune 4.2°N of Moon
17–18		Leonid shower maximum
18	01:51	Uranus 1.5°N of Moon
19	08:57	Full Moon
19	09:27	Partial lunar eclipse (Americas, Pacific, Australia, Asia)
20	12:53	Aldebaran 6.4°S of Moon
21	02:13	Moon at apogee (406,279 km)
24	03:58	Pollux 2.5°N of Moon
26	23:00	Regulus 5.2°S of Moon
27	03:50	Dwarf planet (1) Ceres at opposition (mag. 7.0)
27	12:28	Last Quarter
29	04:39	Mercury at superior conjunction
30	22:35	Spica 5.9°S of Moon

*These objects are close together for an extended

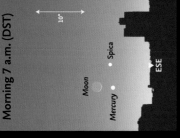

Morning 7 a.m. (DST)

Mercury · Spica · ESE — 10°

November 1 · Before sunrise, Mercury and Spica are low in the east-southeastern sky.

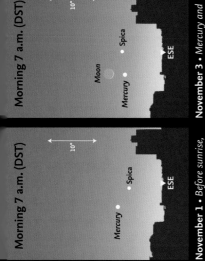

Morning 7 a.m. (DST)

Mercury · Spica · Moon · ESE — 10°

November 3 · Mercury and Spica are joined by the narrow crescent Moon.

Evening 5:30 p.m.

Nunki · 8 · Venus · 7 · SSW · SW — 10°

November 7–8 · The waxing crescent Moon passes Venus and Nunki, shortly afte sunset.

Evening 5:30 p.m.

20 · Nunki · 19 · 18 · Venus · SW — 5°

November 18–20 · Venus passes Nunki, low in the southwestern sky. (Note that the scale of the

Evening 11 p.m.

Alhena · Procyon · Castor · Pollux · Moon · E · 30° — 10°

November 23 · In the eastern sky, the Moon lines up with Castor and Pollux. Procyon and Alhena are farther

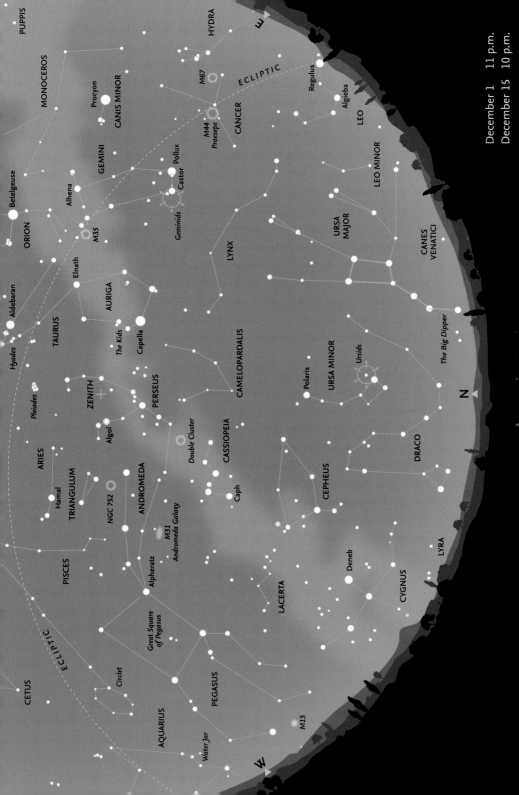

December – Looking North

Ursa Major has now swung around and is starting to "climb" in the east. The fainter stars in the southern part of the constellation are now fully in view. The other bear, *Ursa Minor*, "hangs" below *Polaris* in the north. Directly above it is the faint constellation of *Camelopardalis*, with the other inconspicuous circumpolar constellation, *Lynx*, to its east. *Vega* (α Lyrae) is now below the horizon in the northwest, but *Deneb* (α Cygni) and most of *Cygnus* remain visible farther west. In the east, *Regulus* (α Leonis) and the constellation of *Leo* are beginning to rise above the horizon. *Cancer* stands high in the east, with *Gemini* even higher in the sky. *Perseus* is at the zenith, with *Auriga* and *Capella* between it and Gemini. Because it is so high in the sky, now is a good time to examine the star clouds of the fainter portion of the Milky Way, between *Cassiopeia* in the west to Gemini and *Orion* in the east.

Meteors

There is one significant meteor shower in December (the last major shower of the year). This is the *Geminid* shower, which is visible over the period December 3–16 and comes to maximum on December 14, when the Moon is waxing gibbous so conditions are not wonderful. It is one of the most active showers of the year, and in some years is the strongest, with a peak rate of around 100 meteors per hour. It is the one major shower that shows good activity before midnight. The meteors have been found to have a much higher density than other meteors (which are derived from cometary material). It was eventually established that the Geminids and the asteroid Phaeton had similar orbits. So the Geminids are assumed to consist of denser, rocky material. They are slower than most other meteors and often appear to last longer. The brightest often break up into numerous luminous fragments that follow similar paths across the sky. There is a second shower: the *Ursids*, active December 17–26, peaking on December 22–23,

with a rate at maximum of 5–10, occasionally rising to 25 per hour. Maximum in 2021 is when the Moon is waning gibbous, so conditions for this shower are unfavorable. The parent body is Comet 8P/Tuttle.

The constellation of Cassiopeia is a familiar sight among the northern circumpolar constellations. It is always above the horizon, on the opposite side of Polaris (the North Star) to the equally well-known asterism of the seven stars of the Big Dipper.

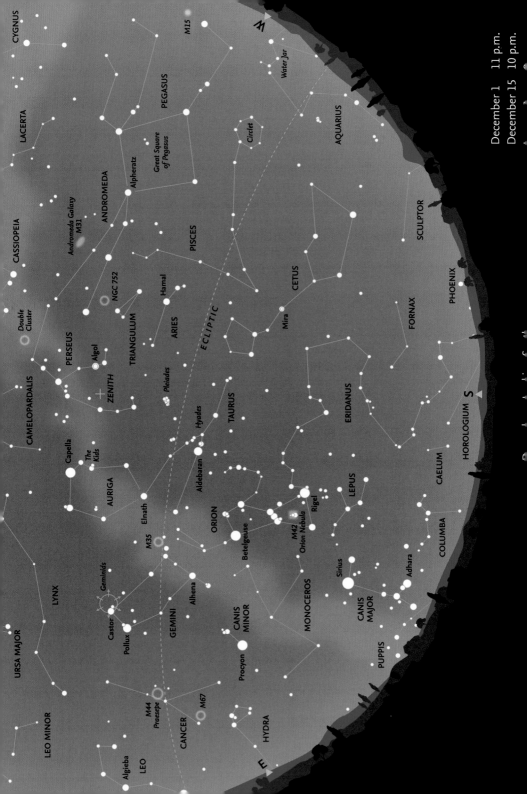

December 1   11 p.m.
December 15   10 p.m.

December – Looking South

The fine open cluster of the *Pleiades* is due south around 10 p.m., high in the sky, with the *Hyades* cluster, *Aldebaran* and the rest of *Taurus* clearly visible to the east. *Auriga* (with *Capella*) and *Gemini* (with *Castor* and *Pollux*) are both well placed for observation. *Orion* has made a welcome return to the winter sky, and both *Canis Minor* (with *Procyon*) and *Canis Major* (with *Sirius*, the brightest star in the sky) are now well above the horizon. The small, poorly known constellation of *Lepus* lies to the south of Orion, with *Columba* closer to the horizon. The northernmost portion of *Eridanus* is clearly visible in the south. In the west, *Aquarius* has now disappeared, and *Cetus* is becoming lower, but *Pisces* is still easily seen, as are the constellations of *Aries*, *Triangulum* and *Andromeda* above it. The Great Square of *Pegasus* is starting to plunge down toward the western horizon, and because of its orientation on the sky appears more like a large diamond, standing on one point, than a square.

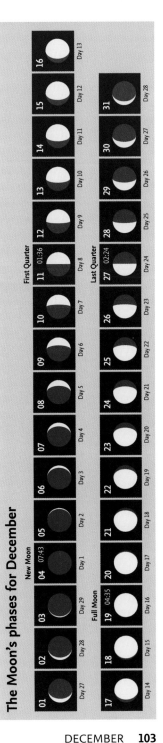

The constellation of Taurus contains two contrasting open clusters: the compact Pleiades, with its striking blue-white stars, and the more scattered, "V"-shaped Hyades, which are much closer to us. Orange Aldebaran (α Tauri) is not related to the Hyades, but lies between it and the Earth.

The Moon's phases for December

New Moon

| 01 Day 27 | 02 Day 28 | 03 Day 29 | 04 07:43 Day 1 | 05 Day 2 | 06 Day 3 | 07 Day 4 | 08 Day 5 | 09 Day 6 | 10 Day 7 | 11 01:36 Day 8 First Quarter | 12 Day 9 | 13 Day 10 | 14 Day 11 | 15 Day 12 | 16 Day 13 |

Full Moon

| 17 Day 14 | 18 Day 15 | 19 04:35 Day 16 | 20 Day 17 | 21 Day 18 | 22 Day 19 | 23 Day 20 | 24 Day 21 | 25 Day 22 | 26 Day 23 | 27 02:24 Day 24 Last Quarter | 28 Day 25 | 29 Day 26 | 30 Day 27 | 31 Day 28 |

December – Moon and Planets

The Moon

On December 4, at New Moon, there is a total solar eclipse, visible from Antarctica only. That day the Moon is at perigee (356,794 km), which is the closest to Earth in 2021, it also passes 3.9° north of *Antares* in *Scorpius*. On December 7 to 9, in the evening sky, the young waxing crescent Moon passes, in succession, *Venus* (December 7), *Saturn* (December 8) and *Jupiter* (December 9). On December 17, nearing Full, the Moon is 6.4° north of *Aldebaran*. On December 21, waning gibbous, it is 3.0° south of *Pollux* in *Gemini*. On December 24, it passes 5.1° north of *Regulus*. On December 31, in the morning sky, the waning crescent Moon is close to *Mars* and *Antares* and passes between the two.

The planets

Mercury is initially too close to the Sun to be visible, where it was actually occulted by the Moon on November 3. Towards the end of the month it moves into the evening twilight and may be glimpsed, low in the sky, when it is close to Venus. *Venus* is in *Sagittarius* but close to the Sun. It may be glimpsed for a short period in the evening sky at the end of the month. *Mars* begins the month in *Libra*, but soon becomes lost near the Sun. *Jupiter* (mag. -2.3 to -2.1) is in *Capricornus* and *Aquarius* moving eastwards, and is visible in the evening sky. *Saturn* (mag. 0.7) is also in *Capricornus*, with slow direct motion, but closer to the horizon than Jupiter. *Uranus* (mag. 5.7) is slowly retrograding in *Aries*. *Neptune* (mag. 7.8) is in *Aquarius* and resumes direct motion on December 4.

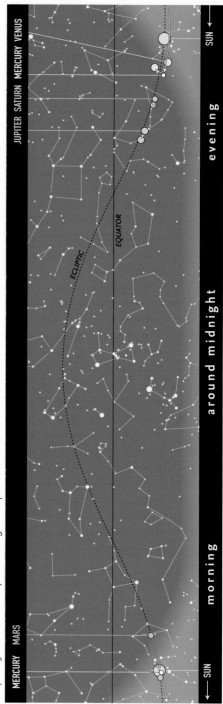

The path of the Sun and the planets along the ecliptic in December.

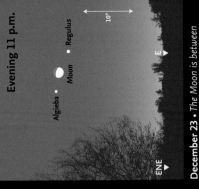

Evening 11 p.m.

Algieba •
Moon • Regulus

ENE
E

10°

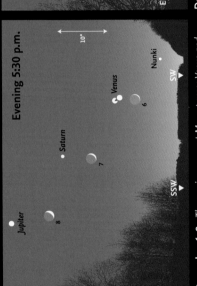

Evening 5:30 p.m.

Jupiter •
Saturn •
8
7
Venus
6
Nunki •

SSW
SW

10°

December 6–8 • *The waxing crescent Moon passes Venus and Saturn, in the south-southwestern sky. One day later (December 9) it has passed Jupiter as well.*

December 23 • *The Moon is between Regulus and Algieba, almost due east.*

Morning 6:30 a.m.

Sabik •
Moon
Mars •
Antares

ESE
SE

10°

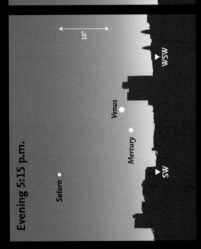

Evening 5:15 p.m.

Saturn •
Mercury •
Venus

SW
WSW

10°

December 29 • *After sunset, Venus and Mercury are close together, low in the southwest. Saturn may be found a little higher and farther south.*

December 31 • *The crescent Moon in the company of Mars and its rival. (The name Antares means "Rival of Mars".) Sabik lines up with Mars and Antares.*

Calendar for December

03–16		Geminid meteor shower
03	00:28	Mars 0.7°S of Moon
04	03:09 ·	Antares 3.9°S of Moon
04	07:33	Total solar eclipse (Antarctica)
04	07:43	New Moon
04	10:04	Moon at perigee (356,794 km, closest of year)
04	12:43	Mercury 0.0°N of Moon
07	00:48	Venus 1.98°N of Moon
08	01:49	Saturn 4.2°N of Moon
09	06:10	Jupiter 4.5°N of Moon
11	00:44	Neptune 4.2°N of Moon
11	01:36	First Quarter
14		Geminid shower maximum
15	05:53	Uranus 1.5°N of Moon
17–26		Ursid meteor shower
17	19:05	Aldebaran 6.4°S of Moon
18	02:15	Moon at apogee (406,320 km)
19	04:35	Full Moon
21	09:56	Pollux 3.0°N of Moon
21	15:59	Winter solstice
22–23		Ursid shower maximum
24	05:14	Regulus 5.1°S of Moon
26	18:00 *	Mars 4.6°N of Antares
27	02:24	Last Quarter
28	07:30	Spica 5.8°S of Moon
29	01:00 *	Mercury 4.2°S of Venus
31	14:24	Antares 3.9°S of Moon
31	20:13	Mars 1.0°N of Moon

* These objects are close together for an extended period around this time.

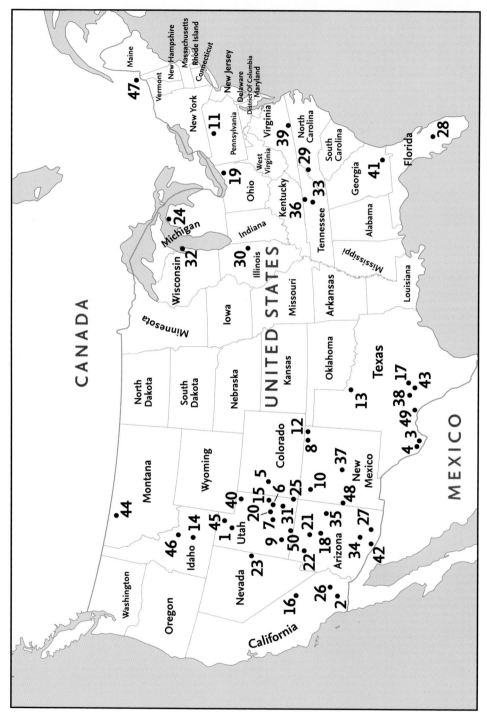

Dark Sky Sites

International Dark-Sky Association Sites

The *International Dark-Sky Association* (IDA) recognises various categories of sites that offer areas where the sky is dark at night, free from light pollution and particularly suitable for astonomical observing. There are numerous sites in North America, shown on the map and listed here.

Details of IDA are at: https://www.darksky.org/. Information on the various categories and individual sites are at: https://www.darksky.org/our-work/conservation/idsp/

Many of these sites have major observatories or other facilities available for public observing (often at specific dates or times).

Parks

1 *Antelope Island State Park* (UT)
2 *Anza-Borrego Desert State Park* (CA)
3 *Big Bend National Park* (TX)
4 *Big Bend Ranch State Park* (TX)
5 *Black Canyon of the Gunnison National Park* (CO)
6 *Canyonlands National Park* (UT)
7 *Capitol Reef National Park* (UT)
8 *Capulin Volcano National Monument* (NM)
9 *Cedar Breaks National Monument* (UT)
10 *Chaco Culture National Historical Park* (NM)
11 *Cherry Springs State Park* (PA)
12 *Clayton Lake State Park* (NM)
13 *Copper Breaks State Park* (TX)
14 *Craters Of The Moon National Monument* (ID)
15 *Dead Horse Point State Park* (UT)
16 *Death Valley National Park* (CA)
17 *Enchanted Rock State Natural Area* (TX)
18 *Flagstaff Area National Monuments* (AZ)
19 *Geauga Observatory Park* (OH)
20 *Goblin Valley State Park* (UT)
21 *Grand Canyon National Park* (AZ)
22 *Grand Canyon-Parashant National Monument* (AZ)
23 *Great Basin National Park* (NV)
24 *Headlands* (MI)
25 *Hovenweep National Monument* (UT)
26 *Joshua Tree National Park* (CA)
27 *Kartchner Caverns State Park* (AZ)
28 *Kissimmee Prairie Preserve State Park* (FL)
29 *Mayland Earth to Sky Park & Bare Dark Sky Observatory* (NC)
30 *Middle Fork River Forest Preserve* (IL)
31 *Natural Bridges National Monument* (UT)
32 *Newport State Park* (WI)
33 *Obed Wild and Scenic River* (TN)
34 *Oracle State Park* (AZ)
35 *Petrified Forest National Park* (AZ)
36 *Pickett CCC Memorial State Park & Pogue Creek Canyon State Natural Area* (TN)
37 *Salinas Pueblo Missions National Monument* (NM)
38 *South Llano River State Park* (TX)
39 *Staunton River State Park* (VA)
40 *Steinaker State Park* (UT)
41 *Stephen C. Foster State Park* (GA)
42 *Tumacácori National Historical Park* (AZ)
43 *UBarU Camp and Retreat Center* (TX)
44 *Waterton-Glacier International Peace Park* (Canada/MT)
45 *Weber County North Fork Park* (UT)

Reserves

46 *Central Idaho* (ID)
47 *Mont-Mégantic* (Québec)

Sanctuaries

48 *Cosmic Campground* (NM)
49 *Devils River State Natural Area – Del Norte Unit* (TX)
50 *Rainbow Bridge National Monument* (UT)

RASC Recognized Dark-Sky Sites

Canadian Dark-Sky Sites

The Royal Astronomical Societ of Canada (RASC) has developed formal guidelines and requirements for three types of light-restricted protected areas: Dark-Sky Preserves, Urban Star Parks and Nocturnal Preserves. The focus of the Canadian Program is primarily to protect the nocturnal environment; therefore, the outdoor lighting requirements are the most stringent, but also the most effective. Canadian Parks and other areas that meet these guidelines and successfully apply for one of these designations are officially recognised. Many parks across Canada have been designated in recent years – see the list below and the RASC website: https://www.rasc.ca/dark-sky-site-designations.

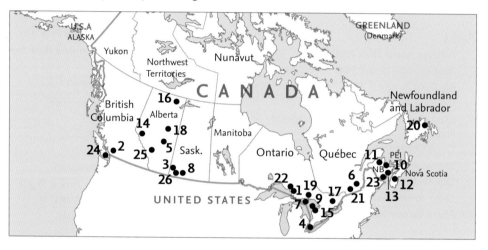

Dark-Sky Preserves

1 *Torrance Barrens Dark-Sky Preserve* (ON)
2 *McDonald Park Dark-Sky Park* (BC)
3 *Cypress Hills Inter-Provincial Park Dark-Sky Preserve* (SK/AB)
4 *Point Pelee National Park* (ON)
5 *Beaver Hills and Elk Island National Park* (AB)
6 *Mont-Mégantic International Dark-Sky Preserve* (QC)
7 *Gordon's Park* (ON)
8 *Grasslands National Park* (SK)
9 *Bruce Peninsula National Park* (ON)
10 *Kouchibouguac National Park* (NB)
11 *Mount Carleton Provincial Park* (NB)
12 *Kejimkujik National Park* (NS)
13 *Fundy National Park* (NB)
14 *Jasper National Park Dark-Sky Preserve* (AB)

15 *Bluewater Outdoor Education Centre – Wiarton* (ON)
16 *Wood Buffalo National Park* (AB)
17 *North Frontenac Township* (ON)
18 *Lakeland Provincial Park and Provincial Recreation Area* (AB)
19 *Killarney Provincial Park* (ON)
20 *Terra Nova National Park* (NL)
21 *Au Diable Vert* (QC)
22 *Lake Superior Provincial Park* (ON)

Urban Star Parks

23 *Irving Nature Park* (NB)
24 *Cattle Point, Victoria* (BC)

Nocturnal Preserves

25 *Ann and Sandy Cross Conservation Area* (AB)
26 *Old Man on His Back Ranch* (SK)

Glossary and Tables

aphelion	The point on an orbit that is farthest from the Sun.
apogee	The point on its orbit at which the Moon is farthest from the Earth.
appulse	The apparently close approach of two celestial objects; two planets, or a planet and star.
astronomical unit	(AU) The mean distance of the Earth from the Sun, 149,597,870 km.
celestial equator	The great circle on the celestial sphere that is in the same plane as the Earth's equator.
celestial sphere	The apparent sphere surrounding the Earth on which all celestial bodies (stars, planets, etc.) seem to be located.
conjunction	The point in time when two celestial objects have the same celestial longitude. In the case of the Sun and a planet, superior conjunction occurs when the planet lies on the far side of the Sun (as seen from Earth). For Mercury and Venus, inferior conjuction occurs when they pass between the Sun and the Earth.
direct motion	Motion from west to east on the sky.
ecliptic	The apparent path of the Sun across the sky throughout the year. Also: the plane of the Earth's orbit in space.
elongation	The point at which an inferior planet has the greatest angular distance from the Sun, as seen from Earth.
equinox	The two points during the year when night and day have equal duration. Also: the points on the sky at which the ecliptic intersects the celestial equator. The vernal (spring) equinox is of particular importance in astronomy.
gibbous	The stage in the sequence of phases at which the illumination of a body lies between half and full. In the case of the Moon, the term is applied to phases between First Quarter and Full, and between Full and Last Quarter.
inferior planet	Either of the planets Mercury or Venus, which have orbits inside that of the Earth.
magnitude	The brightness of a star, planet or other celestial body. It is a logarithmic scale, where larger numbers indicate fainter brightness. A difference of 5 in magnitude indicates a difference of 100 in actual brightness, thus a first-magnitude star is 100 times as bright as one of sixth magnitude.
meridian	The great circle passing through the North and South Poles of a body and the observer's position; or the corresponding great circle on the celestial sphere that passes through the North and South Celestial Poles and also through the observer's zenith.
nadir	The point on the celestial sphere directly beneath the observer's feet, opposite the zenith.
occultation	The disappearance of one celestial body behind another, such as when stars or planets are hidden behind the Moon.
opposition	The point on a superior planet's orbit at which it is directly opposite the Sun in the sky.
perigee	The point on its orbit at which the Moon is closest to the Earth.
perihelion	The point on an orbit that is closest to the Sun.
retrograde motion	Motion from east to west on the sky.
superior planet	A planet that has an orbit outside that of the Earth.
vernal equinox	The point at which the Sun, in its apparent motion along the ecliptic, crosses the celestial equator from south to north. Also known as the First Point of Aries.
zenith	The point directly above the observer's head.
zodiac	A band, stretching 8° on either side of the ecliptic, within which the Moon and planets appear to move. It consists of 12 equal areas, originally named after the constellation that once lay within it.

The Constellations

There are 88 constellations covering the whole of the celestial sphere, but 16 of these in the southern hemisphere can never be seen (even in part) from a latitude of 40°N, so are omitted from this table. The names themselves are expressed in Latin, and the names of stars are frequently given by Greek letters (see next page) followed by the genitive of the constellation name. The genitives and English names of the various constellations are included.

Name	Genitive	Abbr.	English name
Andromeda	Andromeda	And	Andromeda
Antlia	Antliae	Ant	Air Pump
Aquarius	Aquarii	Aqr	Water Bearer
Aquila	Aquilae	Aql	Eagle
Ara	Arae	Ara	Altar
Aries	Arietis	Ari	Ram
Auriga	Aurigae	Aur	Charioteer
Boötes	Boötis	Boo	Herdsman
Caelum	Caeli	Cae	Burin (Chisel)
Camelopardalis	Camelopardalis	Cam	Giraffe
Cancer	Cancri	Cnc	Crab
Canes Venatici	Canum Venaticorum	CVn	Hunting Dogs
Canis Major	Canis Majoris	CMa	Big Dog
Canis Minor	Canis Minoris	CMi	Little Dog
Capricornus	Capricorni	Cap	Sea Goat
Cassiopeia	Cassiopeiae	Cas	Cassiopeia
Centaurus	Centauri	Cen	Centaur
Cepheus	Cephei	Cep	Cepheus
Cetus	Ceti	Cet	Whale
Columba	Columbae	Col	Dove
Coma Berenices	Comae Berenices	Com	Berenice's Hair
Corona Australis	Coronae Australis	CrA	Southern Crown
Corona Borealis	Coronae Borealis	CrB	Northern Crown
Corvus	Corvi	Crv	Crow
Crater	Crateris	Crt	Cup
Cygnus	Cygni	Cyg	Swan
Delphinus	Delphini	Del	Dolphin
Draco	Draconis	Dra	Dragon
Equuleus	Equulei	Equ	Little Horse
Eridanus	Eridani	Eri	River Eridanus
Fornax	Fornacis	For	Furnace
Gemini	Geminorum	Gem	Twins
Grus	Gruis	Gru	Crane
Hercules	Herculis	Her	Hercules
Horologium	Horologii	Hor	(Pendulum) Clock
Hydra	Hydrae	Hya	Water Snake

Name	Genitive	Abbr.	English name
Indus	Indi	Ind	Indian
Lacerta	Lacertae	Lac	Lizard
Leo	Leonis	Leo	Lion
Leo Minor	Leonis Minoris	LMi	Little Lion
Lepus	Leporis	Lep	Hare
Libra	Librae	Lib	Scales
Lupus	Lupi	Lup	Wolf
Lynx	Lyncis	Lyn	Lynx
Lyra	Lyrae	Lyr	Lyre
Microscopium	Microscopii	Mic	Microscope
Monoceros	Monocerotis	Mon	Unicorn
Norma	Normae	Nor	Level (Square)
Ophiuchus	Ophiuchi	Oph	Serpent Bearer
Orion	Orionis	Ori	Orion
Pegasus	Pegasi	Peg	Pegasus
Perseus	Persei	Per	Perseus
Phoenix	Phoenicis	Phe	Phoenix
Pisces	Piscium	Psc	Fishes
Piscis Austrinus	Piscis Austrini	PsA	Southern Fish
Puppis	Puppis	Pup	Stern
Pyxis	Pyxidis	Pyx	Compass
Sagitta	Sagittae	Sge	Arrow
Sagittarius	Sagittarii	Sgr	Archer
Scorpius	Scorpii	Sco	Scorpion
Sculptor	Sculptoris	Scl	Sculptor
Scutum	Scuti	Sct	Shield
Serpens	Serpentis	Ser	Serpent
Sextans	Sextantis	Sex	Sextant
Taurus	Tauri	Tau	Bull
Telescopium	Telescopii	Tel	Telescope
Triangulum	Trianguli	Tri	Triangle
Ursa Major	Ursae Majoris	UMa	Great Bear
Ursa Minor	Ursae Minoris	UMi	Lesser Bear
Vela	Velorum	Vel	Sails
Virgo	Virginis	Vir	Virgin
Vulpecula	Vulpeculae	Vul	Fox

The Greek Alphabet

α	Alpha	ε	Epsilon	ι	Iota	ν	Nu	ρ	Rho	φ (φ)	Phi
β	Beta	ζ	Zeta	κ	Kappa	ξ	Xi	σ (ς)	Sigma	χ	Chi
γ	Gamma	η	Eta	λ	Lambda	ο	Omicron	τ	Tau	ψ	Psi
δ	Delta	θ (ϑ)	Theta	μ	Mu	π	Pi	υ	Upsilon	ω	Omega

Some common asterisms

Belt of Orion	δ, ε, and ζ Orionis
Big Dipper	α, β, γ, δ, ε, ζ and η Ursae Majoris
Cat's Eyes	λ and υ Scorpii
Circlet	γ, θ, ι, λ and κ Piscium
Guards (or Guardians)	β and γ Ursae Minoris
Head of Cetus	α, γ, ξ², μ and λ Ceti
Head of Draco	β, γ, ξ and ν Draconis
Head of Hydra	δ, ε, ζ, η, ρ and σ Hydrae
Keystone	ε, ζ, η and π Herculis
Kids	ε, ζ and η Aurigae
Little Dipper	β, γ, η, ζ, ε, δ and α Ursae Minoris
Lozenge	= Head of Draco
Milk Dipper	ζ, γ, σ, φ and λ Sagittarii
Plough or Big Dipper	α, β, γ, δ, ε, ζ and η Ursae Majoris
Pointers	α and β Ursae Majoris
Sickle	α, η, γ, ζ, μ and ε Leonis
Square of Pegasus	α, β and γ Pegasi with α Andromedae
Sword of Orion	θ and ι Orionis
Teapot	γ, ε, δ, λ, φ, σ, τ and ζ Sagittarii
Wain (or Charles' Wain)	= Big Dipper
Water Jar	γ, η, κ and ζ Aquarii
Y of Aquarius	= Water Jar

Acknowledgements

Denis Buczynski, Portmahomack, Ross-shire – p.32, 35 (Quadrantid fireball), p.65 (Noctilucent clouds)
Steve Edberg, La Cañada, California – pp.35, 43, 89, 91, 95, 101, 103 (constellation photographs)
Akira Fuji, Japan – p.21 (lunar eclipse)
Jens Hackmann, Bad Mergentheim, Germany – p.77 (Perseid fireball)
Bernhard Hubl – pp. 49, 53, 55, 67, 71, 85 (constellation photographs)
Damian Peach – p.29 (comet photos)
peresanz/Shutterstock – p.37 (Orion)
Jan Sandberg – p.24 (Jupiter & satellites)
Ken Sperber, California – p. 83 (Double Cluster)

Specialist editorial support was provided by Dr Emily Drabek-Maunder, Senior Manager of Public Astronomy and Dr Gregory Brown, Public Astronomy Officer at the Royal Observatory, part of Royal Museums Greewich.

Further Information

Books

Bone, Neil (1993), *Observer's Handbook: Meteors*, George Philip, London & Sky Publ. Corp., Cambridge, Mass.
Cook, J., ed. (1999), *The Hatfield Photographic Lunar Atlas*, Springer-Verlag, New York
Dunlop, Storm (2006), *Wild Guide to the Night Sky*, Harper Perennial, New York & Smithsonian Press, Washington D.C.
Dunlop, Storm (2012), *Practical Astronomy*, 2nd edn, Firefly, Buffalo
Dunlop, Storm, Rükl, Antonin & Tirion, Wil (2005), *Collins Atlas of the Night Sky*, HarperCollins, London & Smithsonian Press, Washington D.C.
Grego, Peter (2016), *Moon Observer's Guide*, Firefly, Richmond Hills
Heifetz, Milton D. & Tirion, Wil (2017), *A Walk through the Heavens*, 4th edn, Cambridge University Press, Cambridge
Mellinger, Axel & Hoffmann, Susanne (2005), *The New Atlas of the Stars*, Firefly, Richmond Hills
O'Meara, Stephen J. (2008), *Observing the Night Sky with Binoculars*, Cambridge University Press, Cambridge
Pasachoff, Jay M. (1999), *Peterson Field Guides: Stars and Planets*, 4th edn, Houghton Mifflin, Boston
Ridpath, Ian, ed. (2004), *Norton's Star Atlas*, 20th edn, Pi Press, New York
Ridpath, Ian, ed. (2003), *Oxford Dictionary of Astronomy*, 2nd edn, Oxford University Press, Oxford & New York
Ridpath, Ian & Tirion, Wil (2004), *Collins Gem – Stars*, HarperCollins, London
Ridpath, Ian & Tirion, Wil (2011), *Collins Pocket Guide Stars and Planets*, 4th edn, HarperCollins, London
Ridpath, Ian & Tirion, Wil (2012), *Monthly Sky Guide*, 9th edn, Cambridge University Press, Cambridge
Rükl, Antonín (1990), *Hamlyn Atlas of the Moon*, Hamlyn, London & Astro Media Inc., Milwaukee
Rükl, Antonín (2004), *Atlas of the Moon*, Sky Publishing Corp., Cambridge, Mass.
Scagell, Robin (2015), *Firefly Complete Guide to Stargazing*, Firefly, Richmond Hills
Scagell, Robin & Frydman, David (2014), *Stargazing with Binoculars*, Firefly, Richmond Hills
Scagell, Robin (2014), *Stargazing with a Telescope*, Firefly, Richmond Hills
Sky & Telescope (2017), *Astronomy 2018*, Sky Publishing Corp., Cambridge, Mass.
Tirion, Wil (2011), *Cambridge Star Atlas*, 4th edn, Cambridge University Press, Cambridge
Tirion, Wil & Sinnott, Roger (1999), *Sky Atlas 2000.0*, 2nd edn, Sky Publishing Corp., Cambridge, Mass. & Cambridge University Press, Cambridge

Journals

Astronomy, Astro Media Corp., 21027 Crossroads Circle, P.O. Box 1612, Waukesha, WI 53187-1612.
http://www.astronomy.com
Sky & Telescope, Sky Publishing Corp., Cambridge, MA 02138-1200.
http://www.skyandtelescope.com/

Societies

American Association of Variable Star Observers (AAVSO), 49 Bay State Rd.,
Cambridge, MA 02138.
Although primarily concerned with variable stars, the AAVSO also has a solar section.
American Astronomical Society (AAS), 1667 K Street NW, Suite 800, Washington, DC 20006, New York.
http://aas.org/
American Meteor Society (AMS), Geneseo, New York.
http://www.amsmeteors.org/
Association of Lunar and Planetary Observers (ALPO), ALPO Membership Secretary/Treasurer, P.O. Box 13456, Springfield, IL 62791-3456.
An organization concerned with all forms of amateur astronomical observation, not just the Moon and planets, with numerous coordinated observing sections.
http://alpo-astronomy.org/
Astronomical League (AL), 9201 Ward Parkway Suite #100, Kansas City, MO 64114.
An umbrella organization consisting of over 240 local amateur astronomical societies across the United States.
https://www.astroleague.org/